State and Local Pension Fund Management

PUBLIC ADMINISTRATION AND PUBLIC POLICY

A Comprehensive Publication Program

1. *Public Administration as a Developing Discipline,* Robert T. Golembiewski
2. *Comparative National Policies on Health Care,* Milton I. Roemer, M.D.
3. *Exclusionary Injustice: The Problem of Illegally Obtained Evidence,* Steven R. Schlesinger
5. *Organization Development in Public Administration,* edited by Robert T. Golembiewski and William B. Eddy
7. *Approaches to Planned Change,* Robert T. Golembiewski
8. *Program Evaluation at HEW,* edited by James G. Abert
9. *The States and the Metropolis,* Patricia S. Florestano and Vincent L. Marando
11. *Changing Bureaucracies: Understanding the Organization before Selecting the Approach,* William A. Medina
12. *Handbook on Public Budgeting and Financial Management,* edited by Jack Rabin and Thomas D. Lynch
15. *Handbook on Public Personnel Administration and Labor Relations,* edited by Jack Rabin, Thomas Vocino, W. Bartley Hildreth, and Gerald J. Miller
19. *Handbook of Organization Management,* edited by William B. Eddy
22. *Politics and Administration: Woodrow Wilson and American Public Administration,* edited by Jack Rabin and James S. Bowman
23. *Making and Managing Policy: Formulation, Analysis, Evaluation,* edited by G. Ronald Gilbert
25. *Decision Making in the Public Sector,* edited by Lloyd G. Nigro
26. *Managing Administration,* edited by Jack Rabin, Samuel Humes, and Brian S. Morgan
27. *Public Personnel Update,* edited by Michael Cohen and Robert T. Golembiewski

28. *State and Local Government Administration,* edited by Jack Rabin and Don Dodd
29. *Public Administration: A Bibliographic Guide to the Literature,* Howard E. McCurdy
31. *Handbook of Information Resource Management,* edited by Jack Rabin and Edward M. Jackowski
32. *Public Administration in Developed Democracies: A Comparative Study,* edited by Donald C. Rowat
33. *The Politics of Terrorism: Third Edition,* edited by Michael Stohl
34. *Handbook on Human Services Administration,* edited by Jack Rabin and Marcia B. Steinhauer
36. *Ethics for Bureaucrats: An Essay on Law and Values, Second Edition,* John A. Rohr
37. *The Guide to the Foundations of Public Administration,* Daniel W. Martin
39. *Terrorism and Emergency Management: Policy and Administration,* William L. Waugh, Jr.
40. *Organizational Behavior and Public Management: Second Edition,* Michael L. Vasu, Debra W. Stewart, and G. David Garson
43. *Government Financial Management Theory,* Gerald J. Miller
46. *Handbook of Public Budgeting,* edited by Jack Rabin
49. *Handbook of Court Administration and Management,* edited by Steven W. Hays and Cole Blease Graham, Jr.
50. *Handbook of Comparative Public Budgeting and Financial Management,* edited by Thomas D. Lynch and Lawrence L. Martin
53. *Encyclopedia of Policy Studies: Second Edition,* edited by Stuart S. Nagel
54. *Handbook of Regulation and Administrative Law,* edited by David H. Rosenbloom and Richard D. Schwartz
55. *Handbook of Bureaucracy,* edited by Ali Farazmand
56. *Handbook of Public Sector Labor Relations,* edited by Jack Rabin, Thomas Vocino, W. Bartley Hildreth, and Gerald J. Miller
57. *Practical Public Management,* Robert T. Golembiewski
58. *Handbook of Public Personnel Administration,* edited by Jack Rabin, Thomas Vocino, W. Bartley Hildreth, and Gerald J. Miller
60. *Handbook of Debt Management,* edited by Gerald J. Miller
61. *Public Administration and Law: Second Edition,* David H. Rosenbloom and Rosemary O'Leary
62. *Handbook of Local Government Administration,* edited by John J. Gargan
63. *Handbook of Administrative Communication,* edited by James L. Garnett and Alexander Kouzmin
64. *Public Budgeting and Finance: Fourth Edition,* edited by Robert T. Golembiewski and Jack Rabin
67. *Handbook of Public Finance,* edited by Fred Thompson and Mark T. Green
68. *Organizational Behavior and Public Management: Third Edition,* Michael L. Vasu, Debra W. Stewart, and G. David Garson
69. *Handbook of Economic Development,* edited by Kuotsai Tom Liou
70. *Handbook of Health Administration and Policy,* edited by Anne Osborne Kilpatrick and James A. Johnson

72. *Handbook on Taxation*, edited by W. Bartley Hildreth and James A. Richardson
73. *Handbook of Comparative Public Administration in the Asia-Pacific Basin*, edited by Hoi-kwok Wong and Hon S. Chan
74. *Handbook of Global Environmental Policy and Administration,* edited by Dennis L. Soden and Brent S. Steel
75. *Handbook of State Government Administration,* edited by John J. Gargan
76. *Handbook of Global Legal Policy,* edited by Stuart S. Nagel
78. *Handbook of Global Economic Policy,* edited by Stuart S. Nagel
79. *Handbook of Strategic Management: Second Edition,* edited by Jack Rabin, Gerald J. Miller, and W. Bartley Hildreth
80. *Handbook of Global International Policy,* edited by Stuart S. Nagel
81. *Handbook of Organizational Consultation: Second Edition,* edited by Robert T. Golembiewski
82. *Handbook of Global Political Policy,* edited by Stuart S. Nagel
83. *Handbook of Global Technology Policy,* edited by Stuart S. Nagel
84. *Handbook of Criminal Justice Administration*, edited by M. A. DuPont-Morales, Michael K. Hooper, and Judy H. Schmidt
85. *Labor Relations in the Public Sector: Third Edition*, edited by Richard C. Kearney
86. *Handbook of Administrative Ethics: Second Edition,* edited by Terry L. Cooper
87. *Handbook of Organizational Behavior: Second Edition*, edited by Robert T. Golembiewski
88. *Handbook of Global Social Policy,* edited by Stuart S. Nagel and Amy Robb
89. *Public Administration: A Comparative Perspective, Sixth Edition,* Ferrel Heady
90. *Handbook of Public Quality Management*, edited by Ronald J. Stupak and Peter M. Leitner
91. *Handbook of Public Management Practice and Reform*, edited by Kuotsai Tom Liou
93. *Handbook of Crisis and Emergency Management*, edited by Ali Farazmand
94. *Handbook of Comparative and Development Public Administration: Second Edition,* edited by Ali Farazmand
95. *Financial Planning and Management in Public Organizations,* Alan Walter Steiss and Emeka O. Cyprian Nwagwu
96. *Handbook of International Health Care Systems,* edited by Khi V. Thai, Edward T. Wimberley, and Sharon M. McManus
97. *Handbook of Monetary Policy,* edited by Jack Rabin and Glenn L. Stevens
98. *Handbook of Fiscal Policy,* edited by Jack Rabin and Glenn L. Stevens
99. *Public Administration: An Interdisciplinary Critical Analysis,* edited by Eran Vigoda
100. *Ironies in Organizational Development: Second Edition, Revised and Expanded,* edited by Robert T. Golembiewski
101. *Science and Technology of Terrorism and Counterterrorism*, edited by Tushar K. Ghosh, Mark A. Prelas, Dabir S. Viswanath, and Sudarshan K. Loyalka

102. *Strategic Management for Public and Nonprofit Organizations,* Alan Walter Steiss
103. *Case Studies in Public Budgeting and Financial Management: Second Edition,* edited by Aman Khan and W. Bartley Hildreth
104. *Handbook of Conflict Management,* edited by William J. Pammer, Jr. and Jerri Killian
105. *Chaos Organization and Disaster Management,* Alan Kirschenbaum
106. *Handbook of Gay, Lesbian, Bisexual, and Transgender Administration and Policy,* edited by Wallace Swan
107. *Public Productivity Handbook: Second Edition,* edited by Marc Holzer
108. *Handbook of Developmental Policy Studies,* edited by Gedeon M. Mudacumura, Desta Mebratu and M. Shamsul Haque
109. *Bioterrorism in Medical and Healthcare Administration,* Laure Paquette
110. *International Public Policy and Management: Policy Learning Beyond Regional, Cultural, and Political Boundaries,* edited by David Levi-Faur and Eran Vigoda-Gadot
111. *Handbook of Public Information Systems, Second Edition,* edited by G. David Garson
112. *Handbook of Public Sector Economics,* edited by Donijo Robbins
113. *Handbook of Public Administration and Policy in the European Union,* edited by M. Peter van der Hoek
114. *Nonproliferation Issues for Weapons of Mass Destruction,* Mark A. Prelas and Michael S. Peck
115. *Common Ground, Common Future: Moral Agency in Public Administration, Professions, and Citizenship,* Charles Garofalo and Dean Geuras
116. *Handbook of Organization Theory and Management: The Philosophical Approach, Second Edition,* edited by Thomas D. Lynch and Peter L. Cruise
117. *International Development Governance,* edited by Ahmed Shafiqul Huque and Habib Zafarullah
118. *Sustainable Development Policy and Administration,* edited by Gedeon M. Mudacumura, Desta Mebratu, and M. Shamsul Haque
119. *Public Financial Management,* edited by Howard A. Frank
120. *Handbook of Juvenile Justice: Theory and Practice,* edited by Barbara Sims and Pamela Preston
121. *Emerging Infectious Diseases and the Threat to Occupational Health in the U.S. and Canada,* edited by William Charney
122. *Handbook of Technology Management in Public Administration,* edited by David Greisler and Ronald J. Stupak
123. *Handbook of Decision Making,* edited by Göktuğ Morçöl
124. *Handbook of Public Administration, Third Edition,* edited by Jack Rabin, W. Bartley Hildreth, and Gerald J. Miller
125. *Handbook of Public Policy Analysis,* edited by Frank Fischer, Gerald J. Miller, and Mara S. Sidney
126. *Elements of Effective Governance: Measurement, Accountability and Participation,* edited by Kathe Callahan
127. *American Public Service: Radical Reform and the Merit System,* edited by James S. Bowman and Jonathan P. West
128. *Handbook of Transportation Policy and Administration,* edited by Jeremy Plant

129. *The Art and Practice of Court Administration,* Alexander B. Aikman
130. *Handbook of Globalization, Governance, and Public Administration,* edited by Ali Farazmand and Jack Pinkowski
131. *Handbook of Globalization and the Environment,* edited by Khi V. Thai, Dianne Rahm, and Jerrell D. Coggburn
132. *Personnel Management in Government: Politics and Process, Sixth Edition,* Norma M. Riccucci and Katherine C. Naff
133. *Handbook of Police Administration,* edited by Jim Ruiz and Don Hummer
134. *Handbook of Research Methods in Public Administration, Second Edition,* edited by Kaifeng Yang and Gerald J. Miller
135. *Social and Economic Control of Alcohol: The 21st Amendment in the 21st Century,* edited by Carole L. Jurkiewicz and Murphy J. Painter
136. *Government Public Relations: A Reader,* edited by Mordecai Lee
137. *Handbook of Military Administration,* edited by Jeffrey A. Weber and Johan Eliasson
138. *Disaster Management Handbook,* edited by Jack Pinkowski
139. *Homeland Security Handbook,* edited by Jack Pinkowski
140. *Health Capital and Sustainable Socioeconomic Development,* edited by Patricia A. Cholewka and Mitra M. Motlagh
141. *Handbook of Administrative Reform: An International Perspective,* edited by Jerri Killian and Niklas Eklund
142. *Government Budget Forecasting: Theory and Practice,* edited by Jinping Sun and Thomas D. Lynch
143. *Handbook of Long-Term Care Administration and Policy,* edited by Cynthia Massie Mara and Laura Katz Olson
144. *Handbook of Employee Benefits and Administration,* edited by Christopher G. Reddick and Jerrell D. Coggburn
145. *Business Improvement Districts: Research, Theories, and Controversies,* edited by Göktuğ Morçöl, Lorlene Hoyt, Jack W. Meek, and Ulf Zimmermann
146. *International Handbook of Public Procurement,* edited by Khi V. Thai
147. *State and Local Pension Fund Management,* Jun Peng

Available Electronically

Principles and Practices of Public Administration, edited by Jack Rabin, Robert F. Munzenrider, and Sherrie M. Bartell

PublicADMINISTRATION*netBASE*

State and Local Pension Fund Management

Jun Peng
University of Arizona
Tucson, Arizona, U.S.A.

ROUTLEDGE
Routledge
Taylor & Francis Group
New York London

First published 2009 by Auerbach Publications

Published 2019 by Routledge
711 Third Avenue, New York, NY, 10017
2 Park Square, Milton Park, Abingdon, Oxon OX14 4RN

ISBN 13: 978-0-8493-0548-1 (hbk)

Library of Congress Cataloging-in-Publication Data

Peng, Jun, 1966-
State and local pension fund management / Jun Peng.
p. cm. -- (Public administration and public policy ; 147)
Includes bibliographical references and index.
ISBN 978-0-8493-0548-1
1. United States--Officials and employees--Pensions--Management. 2. Pension trusts--Investments--United States. 3. Local officials and employees--United States--Pensions. I. Title. II. Series.

JK2474.P46 2008
353.5'492130973--dc22 2008008974

Visit the Taylor & Francis Web site at
http://www.taylorandfrancis.com

Contents

Preface

At the peak of the stock market boom in the late 1990s, state and local public pension systems experienced an unprecedented growth in their asset value. As a result, pension benefits for public employees were increased in many states and at the same time government employers reduced their contribution to the pension systems. The stock market downturn between 2000 and 2002, however, brought this almost perfect combination to a halt. All of a sudden, most public pension systems were facing large deficits and the annual pension contribution also skyrocketed for many state and local governments. This all happened at a time when these governments also faced one of the worst fiscal crises since World War II. There has been almost constant media coverage of public pension funding crisis. Such media attention has greatly increased people's awareness of and interest in public pension benefits. Taxpayers, practitioners, researchers, and students in public administration all want to have a good grasp of the issues surrounding public pension management. With $3 trillion and growing in assets under management in 2006, state and local pension systems loom increasingly large over the public finance horizon.

Despite the strong interest in and the need for understanding public pension management, there is a dearth of books dedicated to this important subject. Public financial management textbooks also have very little coverage on this subject. This book seeks to fill the void by providing a basic and systematic discussion of all major issues facing state and local public pension plan management. This discussion consists of two major components. The first is technically oriented and provides a discussion of all the technical details in managing a public pension benefit program, such as pension benefit design, actuarial valuation and funding methods, financial reporting, and pension asset investment management. The discussion of the technical issues is to prepare the reader for the second component, which is more policy oriented. Some major policy issues discussed in this book include: managing public pension programs in the very political context of public budgeting, pension benefit reforms, and the fairness and sustainability of pension benefits in the public sector. Due to the announcement in 2004 of the Governmental Accounting Standards Board (GASB) Statements 43 and 45 that require public employers to determine the liability for promised retiree health benefits, also known as other postemployment benefits (OPEB), the discussion of public retirement benefits in this book also includes OPEB because pension and OPEB benefits are very similar in nature and the unfunded liability related to OPEB in the public sector is substantially larger than that for pension benefits.

With the understanding of major issues in managing retirement benefit programs in the public sector, the reader will have the requisite knowledge to make an informed judgment about public retirement benefit programs and participate in the debate on what reforms, if any, that are needed. This is a major public policy issue that is of concern to not only public employees, but also to every taxpayer in our society.

Acknowledgment

I wish to express my thanks to Jerald Miller of Arizona State University and Dall Forsythe of New York University for encouraging me to write this book and providing suggestions at the beginning of my conceptual thinking, as well as to Dall for providing comments on some of the chapters in the book. I especially want to thank Jack Rabin for acknowledging my public pension research and encouraging me to broaden and deepen it and make it more widely available to both researcher and practitioner audiences.

I also want to thank Laura Anglin, Teri Landin, and Nick Smirensky from the New York State and Local Retirement System for discussing with me the inner workings of a public pension system, from the services provided to actuarial valuation and investment management. A special note of thanks also goes to Buqing Yang of Shanghai University of Economics and Finance who provided assistance in the chapter on actuarial valuation.

Portions of various GASB documents, copyright of the Governmental Accounting Standards Board, 401 Merritt 7, PO Box 5116, Norwalk, CT 06856-5116, are reproduced with permission. Complete copies of these documents are available from the GASB.

Jun Peng

About the Author

Dr. Jun Peng obtained his MPA and doctoral degree in public administration from the University at Albany, State University of New York. He is currently an Associate Professor at the School of Public Administration and Policy, Eller College of Management, University of Arizona. His main teaching and research interest is public financial management in the state and local sector. He has published many journal articles and book chapters on state and local government debt and pension plan management.

Chapter 1

Overview

Due to the confluence of many factors, financial security in retirement has taken on increasing prominence in the media, the public mind, and the national political agenda. These factors include:

- The upcoming retirement of the great baby boomer generation
- The gradual aging of the population in the United States and around the world
- The growing funding gap for the national Social Security and Medicare programs
- The disappearing of traditional pension benefits among workers in the private sector
- The increasing cost of providing retirement benefits to public sector employees
- The high growth rate of medical costs
- The low savings rate among Americans

The effect of these factors is that financial security in retirement has become more uncertain for the current younger generation of workers than for the current retirees or workers who are soon to retire.

This book tells the story of one part of the financial safety net for retirees, namely the retirement benefits provided by state and local governments to their employees. Even though the book is only focused on retirement benefits for public sector employees, it has broader implications for all taxpayers in this country who are inextricably tied in many ways to state and local governments' promises on retirement benefits to their employees.

1.1 Retirement Benefits

Retirement benefits consist of two major types: pension and health. Pension benefits refer to periodic income, or annuity, received by retirees. Health benefits refer to payments for retirees' healthcare premiums and prescription drugs, among other things. Since a pension benefit is by far the larger of the two, this book is mostly focused on issues related to this topic. However, due to the increasing cost related to retiree health benefits in the public sector, a chapter will also be devoted to retiree health benefits toward the end of the book.

In the United States, the analogy for financial security in retirement is commonly referred to as a three-legged stool: Social Security, employer-provided pension benefit, and personal savings, meaning the combined income from these three sources should provide a financially secure retirement. There are at least two levels at which a financially secure retirement can be defined. The first and more basic level of financial security means that income is sufficient to cover the basic living expenses, such as food, clothing, and shelter, in retirement. The second and higher level of financial security means that the income is sufficient to enable the retiree to enjoy a standard of living comparable to that prior to the retirement, meaning the pre- and postretirement income should be comparable. For many retirees, this higher level remains more of a theoretical concept. For lifetime low wage earners, saving for retirement is an extremely difficult task. What makes it worse is that they tend to work in part-time jobs for smaller employers who do not offer any retirement benefits. That makes Social Security the major and, in some cases, the only source of income during their retirement. Among elderly Social Security beneficiaries, 54 percent of married couples and 74 percent of unmarried persons receive 50 percent or more of their income from Social Security; and 21 percent of married couples and about 43 percent of unmarried persons rely on Social Security for 91 percent or more of their income (Social Security Administration, 2007). Since its inception, Social Security has never been designed to replace all or even most of the income earned prior to retirement. Social Security benefits are more closely linked to the basic level than to the higher level of financial security. That is why Social Security has another complement called the Supplemental Security Income, which is to provide a supplement to those whose Social Security benefit is below a minimum amount needed to cover the basic living expenses. In 2007, the minimum amount guaranteed was $623 for an individual and $934 for a couple.

To achieve the second higher level of financial security, employer-provided pension benefit programs play a very important role. Employers can be divided into two major sectors, public and private. Public employers can be further divided into two groups: (1) the federal government and (2) state and local governments. All full-time employees working for public employers and the vast majority of employees working for medium and large private employers receive employer-provided pension benefits. From an employer's perspective, a pension benefit is part of the compensation to an employee, although it is a delayed compensation, in most cases.

Therefore, a pension benefit becomes an important part of the compensation package for employers to attract and keep employees.

Employer-provided pension benefits can be generally divided into two types: defined benefit and defined contribution. In a **defined benefit plan**, the employer guarantees a certain level of pension benefit to the employee based on his/her salary and years of service. While employer-sponsored defined benefit plans require pension contributions from the employers, such plans in the private sector typically do not require contribution from the employees, whereas those in the public sector typically require the employee to contribute to the plan. It is the employer's responsibility to find sufficient resources to pay for the guaranteed benefit. In a **defined contribution program**, the employee sets aside a certain percentage of his wage or salary in a tax-deferred individual account, matched in part or in full by the employer's contribution, which also goes into the individual account. The employee has full control and responsibility over the account, including investment. When he retires, the level of his pension benefit is based on his individual account balance and is no longer the employer's responsibility. The 401(k) plans in the private sector and 403(b) plans in the public and nonprofit sectors are the most common defined contribution plans.

Personal savings that are specifically for retirement purposes are primarily in two forms: individual retirement account (IRA) and annuities purchased from life insurance companies. Due to its tax incentive, the IRA is an important vehicle that encourages individuals to save for their retirement, especially for those who do not have an employer-provided pension benefit.

1.2 Examining the State and Local Pension Plan Management

This book is about one segment of the immense network of retirement programs in the United States — the defined benefit pension plans that are sponsored by state and local governments. While they provide both defined benefit and defined contribution plans, 90 percent of state and local government employees receive pension benefits of the defined benefit type and, therefore, this book is mostly focused on the management of these plans. There are at least four important reasons for examining the state and local government defined benefit plans.

Reason No. 1: State and local government defined benefit pension plans account for a significant portion of the assets accumulated in the United States to fund retirement. Table 1.1 shows total assets in various retirement programs in 2006 outside the Social Security program, corresponding roughly to the retirement programs described above.

Table 1.1 Total Retirement Assets in 2006 (in $Billion)

	Assets	*Percentage (%)*
Annuity with life insurance companies	2363	14.9
Private pension funds	5581	35.2
S&L government retirement funds	2979	18.8
Federal government retirement funds	1142	7.2
Individual retirement accounts	3808	24.0
Total	15,873	100.0

Source: Federal Reserve Flow of Fund Accounts of the United States (2006): Table L. 225 Life Insurance and Pension Funds Reserves, and Table L. 225.i Individual Retirement Accounts (IRAs).

Based on Federal Reserve data, total assets accumulated for retirement benefits stood at close to $16 trillion at the end of 2006.* Of that amount, roughly $3 trillion was held in state and local public pension plans, or close to 20 percent of total retirement assets in the United States. In 2005, about 18 million active workers and retirees were covered under these state and local pension plans.† With about 60 percent of the assets held in corporate stocks (see Figure 1.3 later in this chapter), state and local pension plans collectively hold a significant ownership of corporations in this country and, thus, exert influence on how they are governed. The size of overall public pension plans is further illustrated by the ranking of both public and private pension plans in terms of pension assets under management. Many state and local public pension systems are among the largest pension systems in the United States as well as in the world. According to the annual survey by the *Pension & Investment* in 2006, 8 of the top 10 and 31 of the top 50 largest pension systems in the United States, ranked by total assets, are state and local public pension systems.‡ California Public Employees Retirement System (CalPERS) is by far the largest in the country with about $250 billion in assets under management in 2007. CalPERS was also the fourth largest pension plan in the world in 2007 (Hua, 2007). Given the overall size of state and local pension assets and so many large public pension systems, studying public pension plan management is important, not only to the

* Total assets in individual retirement accounts were $4.232 trillion, of which $424 billion were held as annuity with life insurance companies. Since this amount was already included in the "Annuity with life insurance companies" figure, to avoid double counting, it was subtracted from the individual retirement account figure, resulting in $3.8 trillion.

† U.S. Census Bureau. State and Local Government Employee Retirement Systems Fiscal Year 2005. http://www.census.gov/govs/www/retire05.html (Accessed December 1, 2006).

‡ Pensions & Investments. The Top 200 Pension Funds/Sponsors. http://pionline.com/apps/pbcs.dll/article?AID=/20070122/CHART/101011730/-1/PENSIONFUNDDIRECTORY (Accessed June 1, 2007).

beneficiaries of public pension benefits, but also to anyone who is interested in public policy regarding how financial assets are managed in the public sector.

Reason No. 2: State and local public pension plans are also an important part of the overall economy. This is illustrated, not only by the total financial assets under management, but also by the pension benefit payments made. State and local pension systems collectively made pension payments to beneficiaries to the tune of $156 billion in fiscal year 2005.* This amount accounted for 1.24 percent of the U.S. gross domestic product (GDP) of $11.7 trillion in that year. In 1994, total payment was $58 billion and accounted for only 0.8 percent of the GDP of $7.07 trillion in that year. Thus, as a percentage of GDP, state and local public pension payment increased by about 50 percent in eleven years. With the aging of the population, state and local pension payments are expected to continue to grow and to play an increasingly important role in the U.S. economy.

Reason No. 3: State and local public pension plans are also an increasingly important part of the overall state and local government finance. In 2005, state and local governments collectively contributed about $60 billion to public pension plans, accounting for an important part of the budget for personnel-related expenses.† And, employees paid an additional $32 billion to pension plans. The management of public pension plans has a direct impact on the overall fiscal health of government finance. Increases in pension contribution reduces the amount available for other popular government services, such as education, healthcare, and public safety, and, thus, has the potential to contribute to or exacerbate fiscal pressure on state and local governments. This is most evident after the 2000 to 2002 stock market decline. Pension contributions for many state and local governments increased significantly in subsequent years, further worsening the already deteriorating financial situation facing them due to the slowing economy. In addition, several large public pension systems in states, such as West Virginia, Illinois, and Oklahoma, have significant funding shortages. Because of increasing pension contributions and funding shortages, there have been calls for pension reforms to rein in the cost to the government. Therefore, understanding how pension contribution is determined and what factors will cause it to decrease or increase over time is critical to understanding the linkage between public pension plans and the rest of government finance. In this sense, pension plan management is not only critical to the financial health of the pension plans themselves, but also to the fiscal health of the entire public sector, and public pension plans should be of concern to every taxpayer who eventually stands behind governments' promise to pay for all the employee pension benefits. This linkage has become even more important now that state and local governments have to take into consideration the long-term financial cost of promised health benefits for retirees in addition to pension benefits.

---

* U.S. Census Bureau. State and Local Government Employee Retirement Systems Fiscal Year 2005.

† Ibid.

Reason No. 4: While state and local pension plans share some similarities with their counterparts in the private sector, they are also different in many important ways, from pension benefit protection to regulation and funding policy. Probably the most import difference is the political environment in which public pension systems exist and the ability of public pension sponsors to establish legislation to change pension-funding policy. Therefore, it is important to understand the unique aspects of public pension plan management and how they affect their fiscal health.

1.3 A Brief History of State and Local Public Pension Plans

The history of public pension plans can be characterized by a gradual spread of coverage among different groups of public employees. It is important to know the history of public pension plan development in understanding two unique features of public pension plans: (1) there are different pension plans for employees based on the type of work they do, and (2) some members of public pension plans are not covered by the federal Social Security program.

In this brief discussion, the history of public pension plans is divided into two stages: pre-1941 and post-1941. The first year nationwide data were available on public pension coverage was in 1941 when the U.S. Census Bureau started systematically collecting data on state and local public pension plans in terms of membership, assets, pension contribution, and investment income.

1.3.1 Pre-1940 History

It is not a mere coincidence that the first-ever public pension plan established was one for police officers in New York City in 1857. Since then, many other large cities also started offering disability/pension plans to their police officers and then to firefighters. Legally, it was much easier to justify the establishment of pension benefits for police officers and firefighters, due to the risky and physically demanding nature of their jobs, which also forced them to stop working at a relatively young age (Sterett, 2003).

The second major category of civil servants to be granted pension benefits was school teachers. However, as the rationale to grant pension benefits to police and firefighters did not extend to the teachers, there had been debates whether teachers deserved a pension and it was not until 1894 when New York City established the first pension plan for school teachers (Sterett, 2003). While the early pension plans for teachers were mostly created by the cities, most of them were later merged into statewide pension systems created by the state governments either just for teachers or for all state and local employees within a state. The reason schoolteachers were singled out as a distinct group of employees for pension benefits is that at the time when public pension plans were being developed in the late nineteenth century and early twentieth century, education was by far the largest public service provided in the public sector and teachers accounted for 40 percent of all state and local employees. After public safety, the educational service provided by teachers was the most visible. Despite the vast expansion of state and local government functions, teachers still account for about 40 percent of the state and local government workforce. In 2003, that total workforce was 18 million strong. Of that, 7.4 million were teachers.

As a group, other general state and local government employees were the last to be granted pension benefits. This unique pattern of gradually spreading pension coverage to different groups of state and local public employees when public pension plans were being established has left an indelible mark on the structure of public pension plans. One of the most unique features of public pension plans is that participation in a particular plan depends on the type of employment. Employees can generally be divided into three major and two minor groups based on their employment. The three major groups are: uniformed employees, teachers, and general employees. While uniformed employees primarily refer to police officers and firefighters, in some states, correctional officers are also classified as a separate category of uniformed officers and have their own pension plans. Also, in some states, highway patrol officers have their own pension plans.

Any employee who is not a uniformed officer or a teacher is usually considered a general employee. The two minor types are judges and elected officials. Many feel that judges need their own pension plans because, due to the nature of the job, they typically assume judgeship fairly late in their career and, thus, need special treatment. Elected officials need their own pension plans because many of them are term-limited and stay in office for only a limited period of time. These latter two groups of public employees are much smaller in size compared to the former three groups.

1.3.2 Post-1940 History

By 1941, when the U.S. Census Bureau started its survey of state and local public retirement systems, all states except Idaho had some kind of pension plan. All together, 1.5 million employees were covered by pension plans, accounting for 46 percent of state and local employees.* There was, however, substantial variation in

* The Census Bureau keeps two sets of employment data. One is on the total number of employees, including both full-time and part-time. The other is on the full-time equivalent employees. Because some part-time employees are not eligible for pension benefits, it is more appropriate to calculate the pension coverage ratio on the basis of full-time equivalent employee number. In Table 1.2, pension coverage ratio is calculated on the basis of full-time equivalent employees.

Table 1.2 Percentage of Employees Covered by Retirement Systems (in Thousands)

Year	*Active Members of Pension Systems*	*Full-Time Equivalent Employees*	*Percentage of Employees Covered*
1942	1494	3249	46.0
1952	3021	4012	75.3
1957	3729	4793	77.8
1962	4960	5958	83.2
1967	6465	7455	86.7
1972	8406	9179	91.6
1982	10,144	10,829	93.7
1992	11,998	13,369	89.7
2002	14,124	15,708	89.9
2004	14,181	15,789	89.8

Source: Statistical Abstract of the United States, 1944, 1945, 1953. U.S. Census Bureau State and Local Government Survey of Employee-Retirement Systems Survey, 1957, 1962, 1967, 1972, 1982, 1992, 2002, 2004.

coverage among these states as well as among different types of employees. For types of employees, the pension coverage for teachers was 58 percent in 1941, whereas the coverage among nonschool employees was only 38 percent,* reflecting the fact that pension plans were established much earlier among teachers as a group than other general employees. States in the Northeast had coverage ratios much higher than the national average, whereas many states in the South and West had coverage ratios much lower than the national average. For example, Connecticut, Massachusetts, Pennsylvania, and New York all had coverage ratios above 60 percent. California had the highest overall coverage ratio of 76 percent, with 100 percent of its teachers already covered by pension plans in 1941. Southern states, such as Mississippi, Alabama, and South Carolina, had coverage ratio in the low single digits.

As can be seen from Table 1.2, pension coverage among state and local government employees expanded quickly following 1941. By 1952, three-fourths of public employees were covered by pension plans. By 1972, 90 percent of all public employees were covered. This coverage ratio has remained fairly steady at 90 percent in the public sector since then. This means that by 1972, pension coverage has been extended to everyone who is eligible for a pension. Those who are not in the defined benefit plans, most likely employees in the public institutions of higher education, are covered under defined contribution plans.

One of the important reasons for the quick expansion in the pension coverage for public employees in the 1940s was the creation in 1935 of the Social Security

* U.S. Census Bureau. Statistical Abstracts of the United States 1942.

program at the federal level. The Social Security program provides retirement, disability, and survivor benefits to workers covered by the program. Workers in the private sector were required to join the program. State and local employees, however, were exempt from this requirement, primarily because of Congress's concern over whether the federal government had the constitutional authority to tax state and local governments to fund the program. While debating whether to join the federal program or not, these government units that did not have a pension plan already in place had the added incentive to establish a pension plan. According to a General Accounting Office (1980) report, nearly one-half of large state and local plans were established between 1931 and 1950. In 1950, 1954, and 1956, Congress passed amendments to allow state governments the option to join the Social Security program and most states agreed to do so.

Because state and local governments initially were not allowed to join the Social Security program, but have been given the option to join since the 1950s, and also because many large public pension plans, especially pension plans for teachers, were already established before the introduction of the Social Security program, there exists in the public sector a unique phenomenon where employees in some states are not in the Social Security program. All public employees in Alaska, Colorado, Louisiana, Massachusetts, and Nevada do not belong to the Social Security program. In several other states, such as California, Illinois, and Texas, between 50 to 60 percent of the public employees are not covered by the Social Security and most of them are schoolteachers (Munnell, 2005). Altogether about 5 million public sector employees are not covered by Social Security. The effect of whether an employee is covered by the Social Security or not can be found in pension benefit design, and will be discussed in Chapter 2.

1.4 Public and Private Sector Pension Benefits

Other unique aspects of public pension benefits can be appreciated only through comparison with pension benefits provided in the private sector. First, the administration of defined benefit pension plans in the private sector is regulated by the federal Employee Retirement Income Security Act (ERISA), whereas public sector DB (defined benefit) pension plans are exempt from ERISA. Whether such exemption has an impact on how public pension plans are managed is discussed in Chapter 4.

Second, as can be seen from the previous section, the DB pension plan coverage among public employees increased gradually to 90 percent in the early 1970s and has remained at that percentage since. This is very different from the historical trend of defined benefit plan coverage among private sector employees. In the private sector, there has been a steady decline in the percentage of employees covered by defined benefit pension plans. In 1980, 84 percent of full-time employees participated in defined benefit plans at medium and large companies (with 100 or more employees) and 20 percent of employees did so at small companies (with fewer than 100 employees) in 1990 (U.S. Department of Labor, 1981, 1991). In 1999, only 42 percent of employees at medium and large companies and 10 percent of employees at small companies were covered by defined benefit pension plans.* By 2006, this percentage continued to decline to 33 percent at median and large companies and down to 9 percent at small companies.† At the same time, the participation rate in defined contribution program increased from 42 percent in 1999 to 70 percent in 2006 at medium and large companies, and from 34 percent in 1999 to 41 percent in 2006 at small companies.‡ The total number of defined benefit pension plans decreased from a peak of 112,000 in the mid-1980s to about 30,000 in the early 2000s, and the total number of active participants in defined benefits pension plan decreased from 22 million in 1985 to 17 million in 2002 (Pension Benefit Guarantee Corporation, 2005). As more and more medium and large companies have frozen their defined benefit plans, the participation in defined benefit plans is expected to continue to decrease in the private sector.

Due to this sharp difference between the public and private sectors in terms of the types of pension benefits being offered, there have been many calls for a switch from defined benefit to defined contribution pension plans in the public sector. This topic will be discussed in detail in Chapter 7.

1.5 State and Local Pension Systems

In this section, we take a closer look at some other important characteristics of state and local pension plans.

1.5.1 Public Pension Systems, Plans, and Funds

When discussing public pension plan management, terminology can be quite confusing. Many times, pension systems, pension plans, and pension funds are used interchangeably. In its most strict technical sense, these three terms have very different meanings.

* U.S. Department of Labor, Bureau of Labor Statistics. "National Compensation Survey: Employee Benefits in Private Industry in the United States, 1999." Supplementary Tables.

† U.S. Department of Labor, Bureau of Labor Statistics. "National Compensation Survey: Employee Benefits in Private Industry in the United States, March 2006," released in August, 2006.

‡ Ibid.

1. **Pension plan**: This refers to a pension program that offers a set of benefits to any eligible participants in this program. A pension plan contains at least three important attributes: who the participants are, what pension benefits are to be offered and at what level, and who should be responsible for the financing of these benefits. A pension plan should also have a plan sponsor. In the public sector, the plan sponsor is the government entity that has created the pension plan, such as a state or local general-purpose government, a school district, or a special district. As discussed earlier, due to historical reasons, a government sponsor can offer several pension plans, each for a different group of employees with a different set of benefits.
2. **Pension fund**: In its strictest sense, pension fund is an accounting concept, referring to an account that holds all the assets accumulated for a particular pension plan. A pension fund is set up as an irrevocable trust, meaning the assets in the fund can only be used for the benefit of the members in the pension plan. Each pension fund then should correspond to one pension plan. Because of this, a pension plan for a group of employees can also be identified officially as a pension (trust) fund.
3. **Pension system**: This refers to the legal fiduciary entity created by the pension plan sponsor to administer the pension plan(s). In most cases, the government entity that creates the pension plan does not administer the pension plan. Instead, it establishes another entity called the public employee retirement system to administer it. This retirement system is governed by a board of trustees who are either appointed by the plan sponsor or elected by plan members, both active and retired. At the local level, a retirement system typically administers just one pension plan. In this case, the pension system can also be called a pension plan as in the case of Los Angeles Water Power Employees Retirement Plan, or a pension (trust) fund as in the case of Marin County (California) Employees Retirement Fund. At the state level, a pension system is more likely to administer multiple pension plans and, thus, is more likely to be identified as a public employee retirement system. Another term that is more commonly used at the local level than at the state level to identify a pension system is "retirement association," as in the case of Santa Barbara County (California) Employees Retirement Association.

In this book, pension fund is used strictly as an accounting concept and pension plan is used to refer to a government pension program. Pension system, however, contains two different meanings in this book. First, it means the legal entity established to administer the pension plan(s). Second, it also refers to multiple pension plans administered by a pension system. The reason pension system takes on these two different meanings is that state-level pension plans are the focus of this book and a state-level pension system is most likely to administer more than one pension plan. In other words, pension plan and public employee retirement system are used interchangeably in this book.

1.5.2 Types of Public Pension Plans

While public pension plans can be categorized based on the employees being covered, they can also be divided into various types based on employer sponsorship: (1) single-employer, (2) agent multiemployer, or (3) cost-sharing multiemployer plan.

1. **Single-employer plan**: This is a pension plan that covers the current and former employees of only one employer. For example, a county or municipal government establishes a plan that only the county or the city's employees participate in. In this case, the county or the city government is fully responsible for the financing of the pension plan. Most of the pension plans at the local level are single-employer plans.
2. **Agent multiemployer plan (or agent plan)**: This is an aggregation of single-employer plans. The individual plans and assets in each plan are pooled together for administrative and investment purposes. However, each plan is completely separate from other plans in terms of benefits calculation, asset accumulation, and financing cost. In other words, separate accounts are maintained for each employer so that the employer's contributions provide benefits only for the employees of that employer. The main purpose of an agent plan is to reduce the administrative and investment cost of a pension plan. This is especially beneficial for smaller pension plans because some parts of the administrative and investment costs are fixed, thus making it more expensive for them.
3. **Cost-sharing multiemployer plan**: This is a single plan with pooling (cost-sharing) arrangements for the participating employers. All funding and investment risks, rewards, and costs, including benefit costs, are shared and are not attributed individually to the employers. In an agent plan, each employer can offer different pension benefits to their employees due to the separation of accounts. In a cost-sharing plan, no separate account exists for each employer. All employers have to offer the same pension benefits to their employees. When joining a cost-sharing plan, the employer loses the independence in designing pension benefits. In the public sector, agent plans are relatively rare and cost-sharing plans are fairly common. Many state-level pension plans are cost-sharing plans that can include hundreds of local level employers in addition to the state government employer. For example, Arizona State Retirement Plan is a cost-sharing plan with 598 participating employers in 2005, such as school districts, cities, towns, counties, special districts, and public authorities. The main purpose of a cost-sharing plan is to realize the economy of scale. Many public employers are too small to manage their own pension plans. By having a large state or county government manage the pension plans of numerous small- to medium-sized employers, significant savings in administrative and investment costs can be realized. Another important benefit of a cost-sharing pension plan is portability of pension benefits within a

state. There will be no change to an employee's pension benefit if she changes jobs in the public sector within the same state, if all state and local government employers participate in one cost-sharing pension plan. Local government employers, however, are not required to join a state-level cost-sharing plan. Larger local government employers typically opt out of the state cost-sharing plan in favor of establishing their own plans so that they can have the flexibility of designing their own benefits. For example, Pima County and Tucson City in Arizona have their own separate pension plans.

A cost-sharing plan is a very important feature of state and local pension plans and is the major reason why so many of the largest pension systems in the United States, in terms of assets, belong to the public sector.

1.5.3 Historical Data on the Number of Pension Systems

Since, in the United States, there are more than 80,000 state and local government units, plus numerous independent government agencies, such as public authorities, it is theoretically possible to have tens of thousands of pension plans and pension systems. The existence of state-level cost-sharing plans, however, cuts the potential number of such plans and systems by a significant margin. While many large counties and municipalities establish pension systems to manage their own pension plans, most local government units are too small to do so and, therefore, join the statewide pension systems. Table 1.3 illustrates the number of pension systems for selected years since 1957.

This 50-year history can be divided into three phases. In the first phase, between 1957 and 1977, there was a major increase in the number of pension systems at both the state and local levels, indicating an increase in more and more pension plans

Table 1.3 Number of State and Local Public Pension Systems

Year	*Total*	*State*	*Local*
1957	2205	147	2058
1972	2304	176	2128
1977	3075	197	2878
1997	2276	212	2064
2000	2208	220	1988
2004	2659	220	2439
2005	2656	222	2434

Source: U.S. Census Bureau State and Local Government Employee-Retirement Systems Survey, 1957, 1972, 1977, 1997, 2000, 2004, and 2005.

being established, consistent with the data on pension coverage shown in Table 1.2. The second phase, between 1977 and 1997, showed some consolidation. There was a slight increase in the number of state-level pension systems, but a significant decrease in local-level pension systems, indicating a merge of local pension systems into state pension systems. Since then, more local pension systems have been created, whereas the number of state-level pension systems has stabilized.

While most states have a comparable number of state-level pension systems, there is far greater variation in the number of local pension systems for each state. It ranges anywhere from no local pension system in some states to several hundred in others. Six states have no local pension systems: Hawaii, Maine, Montana, Nevada, New Mexico, and Utah, meaning all local government employees belong to one or more state-level pension systems. At the other end, six states, with a total of 1808 pension systems, accounted for three-fourths of the 2439 local pension systems in 2002: Florida (157), Illinois (365), Massachusetts (86), Michigan (135), Minnesota (137), and Pennsylvania (928) (U.S. Census Bureau, 2004). Pennsylvania is an outlier in this respect by far, accounting for almost 40 percent of all local systems. Two factors can help explain to some extent such variation among states. First is the number of general-purpose local government units. Illinois, Minnesota, and Pennsylvania were the top three states with most general-purpose local government units in 2002 (U.S. Census Bureau, 2002). The second factor can be the difference in each state's legislation enabling the establishment of local pension plans. In Pennsylvania's case, the decentralized public pension system was largely a result of 50 state statutes establishing local pension plans over a 60-year period (Pennsylvania Public Employee Retirement Commission, 1992). Because of the decentralized pension system structure, three of these six states, Massachusetts, Minnesota, and Pennsylvania, created permanent state-level pension oversight commissions to monitor local pension systems.

In terms of the types of employees covered, most of the pension systems at the local level are for police officers and firefighters. For example, in the Census Survey of 1987 (U.S. Census Bureau, 1990), there were 2213 local pension systems. Of these, 1699 were for limited employee coverage, most of which were for police and firefighters only: 712 systems for police officers only, 652 for firefighters only, and another 200 for both police officers and firefighters.

1.5.4 Breakdown between State- and Local-Level Pension Systems

By combining geographic location and types of employee coverage, the conclusion can be drawn that despite the large number of local pension systems, most are in a few states and they tend to cover a relatively small number of members. It is no wonder that the vast majority of pension assets, members, and beneficiaries belong to state-level pension systems, despite a much smaller number of state-level

Table 1.4 State and Local Pension Systems Comparison in 2004 (in $Million)

		Amount	*Percentage (%)*
Assets			
	Total	2495	100
	State	2079	83
	Local	416	17
Membership			
	Total	17.9	100
	State	16.1	90
	Local	1.8	10
Beneficiaries			
	Total	6.7	100
	State	5.6	84
	Local	1.1	16

Source: U.S. Census Bureau State and Local Government Employee-Retirement Systems Survey, 2004.

pension systems. Table 1.4 shows the distribution of assets, membership, and beneficiaries between state and local pension systems in 2004. State pension systems were responsible for well over 80 percent in each of the three categories, with the asset share corresponding roughly to membership and beneficiary shares. In 1957, only about 63 percent of assets were managed by state-level pension systems. By 1977, 77 percent of assets belonged to state pension systems. This historical trend in the concentration of assets under state pension systems' management mirrored the consolidation of pension systems shown in Table 1.3.

Even within local-level pension systems, there is also a high level of concentration of assets. Of the $383 billion in assets that belonged to local pension systems in 2002, about 44 percent were in pension systems established by the New York City government, Los Angeles county and city governments, San Francisco city and county government, and Chicago city government (U.S. Census Bureau, 2004). New York City alone accounted for 25 percent of all assets held in local pension systems in 2002.

At the state level, most of the assets and members belong to state pension systems that cover broad categories of employees, such as general employers or teachers or both. Many state pension systems that cover narrower categories of employees, such as judges, legislators, and patrol officers, are much smaller in terms of both assets and members. Since 1982, the Wisconsin State Legislature has conducted a biannual survey of 83 state-level pension systems.* These 83 tables can be found in Appendix A at the end of this chapter. According to the U.S. Census of 2002, these 83 state systems had $1.62 trillion in assets under management and 13.8 million members, accounting for 80 percent of total state pension system assets and 86 percent of state pension system membership.

Because of the dominance of state-level pension systems, especially large state-level pension systems that cover broad categories of employees, the discussion of public pension plan management in this book is mostly focused on these state-level pension systems. More specifically, the results of surveys by the Wisconsin State Legislature are the basis for understanding the features of public pension benefit design in Chapter 2 and actuarial valuation in Chapter 3. Two important observations can be drawn from Appendix A and the 2002 Census Bureau survey of state and local pension systems with regard to teacher and local employee coverage at the state level:

1. With regard to teachers, there are 27 states that have a separate statewide plan exclusively for teachers and in the other 23 states, teachers are part of a statewide plan that also covers other state and local employees. Not all local school districts participate in statewide pension plans, exclusively for teachers or otherwise. There were 14 school districts that had their separate pension plans in 2002 with a total membership of about 85,000, a small fraction of all teachers in that year (U.S. Census Bureau, 2004). The Chicago school district had the largest independent pension plan with about $10 billion in assets and about 40,000 members in 2002. Other cities that have large independent school district pension plans are Denver, Colorado; Minneapolis and St. Paul, Minnesota; Kansas City and St. Louis, Missouri; and Omaha, Nebraska.
2. With regard to local employee coverage, there are 37 states that offer the same pension plan to both state employees and local employees and the plan is managed by a state-level retirement system. In 12 states where state pension plans do not cover local employees, there is a statewide pension plan or municipal retirement system just for local governments: Connecticut Municipal Employees Retirement System, Delaware County and Municipal Other Employees Pension Plan, Georgia Municipal Employee Benefit System, Kentucky County Employees Retirement System, Illinois Municipal Retirement Fund, Louisiana Municipal Employees Retirement System, Michigan Municipal Employees Retirement System, Minnesota Public Employees Retirement Association, Missouri Local Government Employees Retirement System, North Carolina Local Government Employees Retirement System, Penn-

* The author would like to thank William Ford of the Wisconsin Legislative Council for providing a copy of all the surveys since 1982. The survey is conducted by the staff of the Wisconsin Retirement Research Committee of the Wisconsin Legislative Council every two years since 1982. The reports on the surveys of 2000, 2002, and 2004 are available at the Wisconsin Legislative Council's Web site at http://www.legis.state.wi.us/lc/publications/bysubject.asp.

sylvania Municipal Retirement System, and Vermont Municipal Employees Retirement System. Massachusetts is the only state that does not have either a state pension plan for local employees or a state-level municipal retirement system.

1.6 Growth of Public Pension System Assets

In this section, we examine the growth of public pension systems in terms of accumulated assets, the source of such growth, and its effect on the long-term fiscal health of pension systems.

1.6.1 Pension Asset Growth

Along with the increase in pension plan membership, assets in pension plans also witnessed tremendous increase over the same period. In 1945, total assets held in pension plans were about $2 billion, indicating that at least some pension plans had already started accumulating assets to pay for future pension benefits by then. Figure 1.1 shows the historical growth of state and local pension assets between 1945 and 2006.

An important characteristic of growth of public pension assets is that it is not geometric, but rather exponential. It took about 30 years for pension assets to reach $100 billion in 1975. It took the next 18 years to go from $100 billion to $1 trillion in 1993. However, it took only 13 years to go from $1 trillion to $3 trillion in

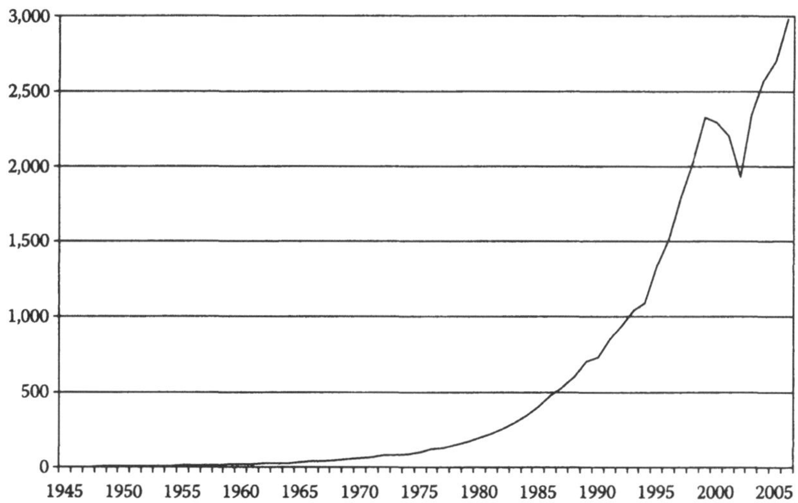

Figure 1.1 State and local pension asset growth, 1945 to 2006 (in $billion). (From U.S. Federal Reserve Bank Flow of Funds, various years.)

2006. Except for a brief period between 2000 and 2003, the pension asset value has been on a continuous upward trajectory. This exponential growth of public pension plans is not only due to the power of compounding of investment return, but also due to the higher average return in the later part of this period than in the earlier part, as can be seen in the next part of this section.

The importance of asset accumulation has to do with the way pension benefits are financed. In pension terminology, this is called funding policy, a topic discussed more fully in Chapter 4. All large public pension plans pursue a funding policy of paying, fully or partially, for the cost of pension benefits when they are being earned by an employee, not when they are being paid to a retiree. Since pension benefits will not be paid out for many years to come, such a funding policy will naturally lead to an increase in pension assets.

1.6.2 Source of Pension Asset Growth

Public pension asset growth comes from three different sources: employee pension contribution, employer pension contribution, and investment return on these contributions. If the total of these three sources is greater than the benefit payment in that year, pension asset value will increase. Due to the power of compounding, investment income becomes a more and more important source of asset growth over time. This can be seen in Table 1.5, which shows the employer and employee pension contributions made and investment income earned for selected years between 1942 and 1986 and for all years between 1991 and 2005.

In the early years of public pension plans, total pension contributions were much greater than investment income. In other words, the pension asset growth in early years was largely due to pension contribution, rather than investment income. It wasn't until 1985 when the annual investment return was nearly equal to the combined employer and employee contributions. The gap between the two sources became increasingly greater since then. This trend, however, was briefly interrupted between 2001 and 2003 when the stock market suffered a major downturn. The substantial reduction in investment income in 2001 was the main reason for the first ever year-to-year decrease in total pension assets, as can be seen in Figure 1.1. This trend can also be viewed from a cumulative perspective. Over this period between 1942 and 2005, it was not until 1994 when the cumulative investment income finally surpassed cumulative pension contributions. By 2005, the total investment income reached $2.4 trillion, about $900 billion more than the total pension contributions made.

Given the importance of investment income, which is based on the rate of return, it is useful to examine the historical investment return trend for public pension funds. The information can be found by dividing the year's investment income into the previous year's total assets. Figure 1.2 shows the annual investment return between 1953 and 2005.

Table 1.5 Total Receipts for Public Pension Systems (in $Million)

		Pension Contribution		
Year	*Total*	*Employee*	*Employer*	*Earnings*
1942	119	65	54	36
1950	539	260	278	70
1960	1819	802	1017	398
1970	7388	2788	4600	2460
1975	13,604	4488	9116	5294
1985	36,878	9479	27,399	34,852
1986	39,185	10,586	28,599	48,965
1990	46,431	13,853	32,578	64,907
1991	49,401	16,238	33,163	58,808
1992	49,582	16,028	33,554	77,222
1993	51,130	16,138	34,992	74,813
1994	54,114	17,341	36,772	79,180
1995	59,611	18,600	41,011	89,232
1996	60,895	19,372	41,523	129,562
1997	65,833	20,931	44,902	161,223
1998	63,685	21,835	41,850	197,631
1999	65,300	23,566	41,724	197,865
2000	65,150	24,994	40,155	231,900
2001	65,282	26,438	38,845	57,941
2002	66,336	27,544	38,792	(72,457)
2003	75,056	28,844	46,212	72,691
2004	91,782	30,786	60,996	315,554
2005	91,283	31,536	59,747	262,178

Source: U.S. Census Bureau State and Local Government Employee-Retirement Systems Survey.

The rate of return was not constant over this period. Except for a brief period in the early 2000s, the investment return has generally been positive. What is most significant is that the average return since the mid-1980s was much higher than the average return in the period prior to that. Therefore, the exponential growth in pension assets seen in Figure 1.1 was not only due to the power of compounding, but also due to the much higher average return since the mid-1980s. A major factor behind this increase was the gradual shift from safe investments, such as fixed-income security, to more risky investments, such as equity, throughout this period.

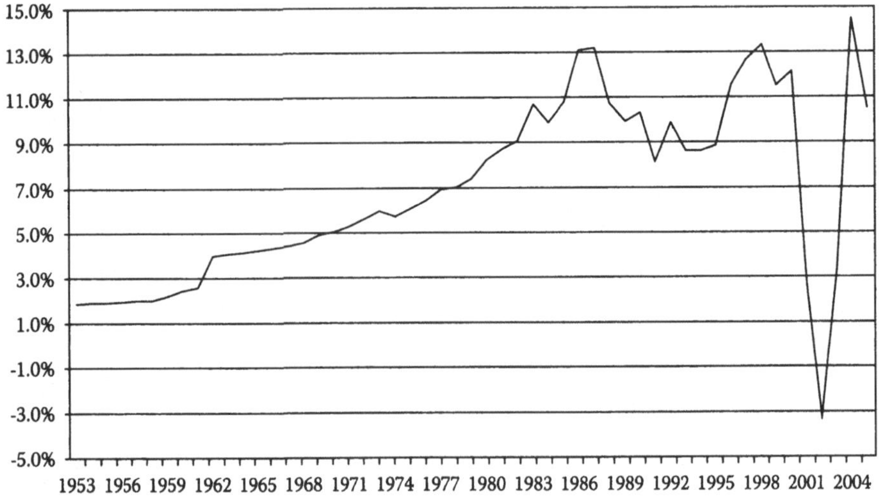

Figure 1.2 Historical return on public pension asset investment. (From U.S. Census Bureau State and Local Employee-Retirement Systems Survey, various years.)

As risky assets in general are associated with higher return, this shift in asset allocation also results in higher return over time. This gradual shift in asset allocation can be seen in Figure 1.3.*

Until the mid 1950s, investment in equity was almost nonexistent and pension assets were invested exclusively in fixed-income securities. Since the mid-1950s, there has been an almost constant increase in asset allocation to equity investment, accompanied by an equally constant decrease in asset allocation to fixed-income investment. While the increase in allocation to equity investment has increased investment return, it has increased volatility in investment return as well. This is clearly shown by the only negative return in the early 2000s over this entire period when overall asset allocation to equity reached 60 percent. This relationship between investment return and risk is the most important aspect of pension asset investment management. Because of the importance of investment income to public pension systems, as shown in this section, Chapter 5 is devoted entirely to investment management.

Table 1.5 also shows the important relationship between investment income and pension contribution, especially employer pension contribution. Throughout this entire period, employee contribution has shown a steady increase. This is not surprising as the employee contribution rate is typically set to a certain percentage of

* The allocation to fixed income and equity does not always add up to 100 percent, especially since the mid 1980s. This is because while these two types of assets account for the lion's share of the investment portfolio, other assets include cash and alternative investments, a topic discussed in Chapter 5. Since the early 2000s, there has been a gradual shift from traditional investments, such as fixed income and equity, to alternative investments.

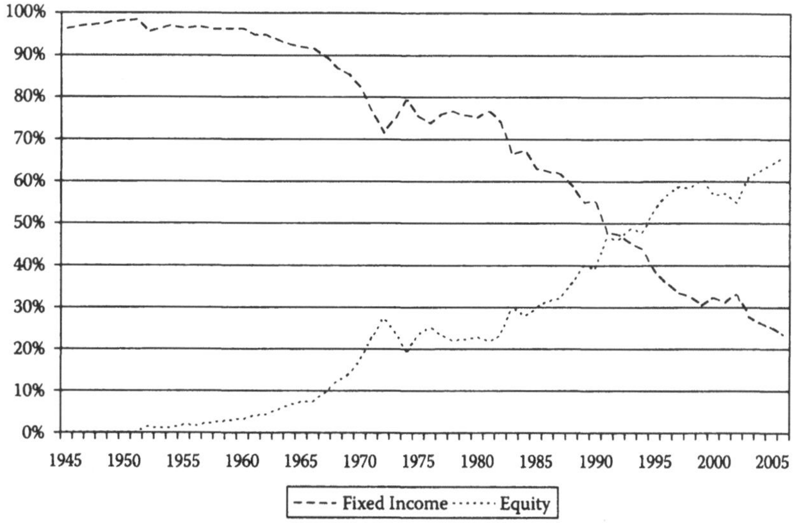

Figure 1.3 Historical asset allocation for fixed income and equity. (From U.S. Federal Reserve Flow of Funds, various years.)

an employee's salary and this percentage rarely changes. Since salary goes up every year due to inflation and merit increase, the employee contribution amount also goes up. The same cannot be said of employer contribution. Employer contribution showed a steady increase until 1997. Between 1997 and 2002, annual employer contribution showed a steady decrease of about 13 percent, largely as a result of the substantial investment income earned by the pension systems. However, when the investment return turned lower in 2001 and 2003, and even negative in 2002, employer pension contribution went up again. Between 2001 and 2004, annual employer pension contribution increased by 57 percent. This increase stabilized in 2005 as a result of large investment income in 2004. This inverse relationship between investment income and employer contribution is one of the most important aspects of public pension plan management. How to manage this relationship is the focus of Chapter 6. Despite the growing importance of investment income over time, pension contribution still plays a critical role, as it is the original source for reaping greater investment income in the future.

1.6.3 Sufficiency of Pension Asset Growth

While the increase in pension assets certainly indicates that state and local governments have accumulated more assets over time to pay for pension benefits, the increase itself does not indicate whether state and local pension plans have accumulated sufficient assets at any given time. What matters most is whether the pension assets accumulated are greater or less than the projected pension benefits (or pension liabilities from the government employer's perspective) to be paid at a particular point in time. If pension assets are less than pension liabilities, the difference is called *unfunded pension liability* and the pension plan is considered underfunded. How well a pension plan is funded is measured by a funding ratio, which is calculated by dividing pension assets into pension liabilities.* A funding ratio less than 1 means the pension plan is underfunded. Up until the early 1990s, concerns were raised over the fact that not enough assets had been accumulated in public pension plans to offset the pension liabilities. The first systematic analysis of the public plans' financial condition was conducted in 1978 by the U.S. House of Representatives' Pension Task Force. It found that state and local pension plans had unfunded pension liabilities between $150 billion and $175 billion in 1975 and the funding ratio was about 50 percent, meaning the assets accumulated were only half of the projected pension liabilities.† Since the mid-1970s, however, the funding situations of public pension plans have improved gradually. By 1992, the overall funding ratio of public plans had increased to 82 percent, according to a Public Pension Coordinating Council survey and the total unfunded pension liabilities in 1992 stood at $200 billion, about half of the unfunded liability level in 1975, which would be $400 billion if measured in 1992 dollars (General Accounting Office, 1996).

With the strong stock market return throughout the 1990s and the increasing asset allocation to equity investment by public pension systems, the pension asset value increased at a much faster pace than pension liability, leading to a substantial improvement in the overall funding ratio, as can be seen in Figure 1.4.‡

By 2000, the average funding ratio for public pension plans reached 100 percent for the first time in history and many pension plans were overfunded by a large margin. This overfunding led many state and local governments to increase pension benefits and cut pension contributions, as can be seen from Table 1.5. The stock market decline between 2000 and 2003 finally ended the steady improvement in the funding ratio of public pension plans since the mid-1970s. By the end of fiscal year 2005, the average funding ratio dropped to 84 percent. Only a few large pension plans still had a funding ratio above 100 percent in 2005. This drop in funding ratio also contributed to the substantial increase in employer contribution, as seen in Table 1.5. This quick rise and fall in the pension-funding ratio between the early 1990s and the early 2000s hold many lessons for managing public pension plans in the future. How to manage public pension plans in the face of ups and downs in the financial market is a major theme in Chapter 6.

---

* More precise definition of unfunded pension liability is discussed in Chapter 3.

† As it will be explained more fully in Chapter 3, it is not uncommon for pension funds, public or private, to be underfunded by a significant amount at an early stage after their establishment, as many of them do not have sufficient time to reduce the liability related to services provided before the establishment of the pension plan.

‡ The National Association of State Retirement Administrators has given the author the permission to use the data used in this figure. The data combines the findings of NASRA's Public Fund Survey and Standard & Poor's research.

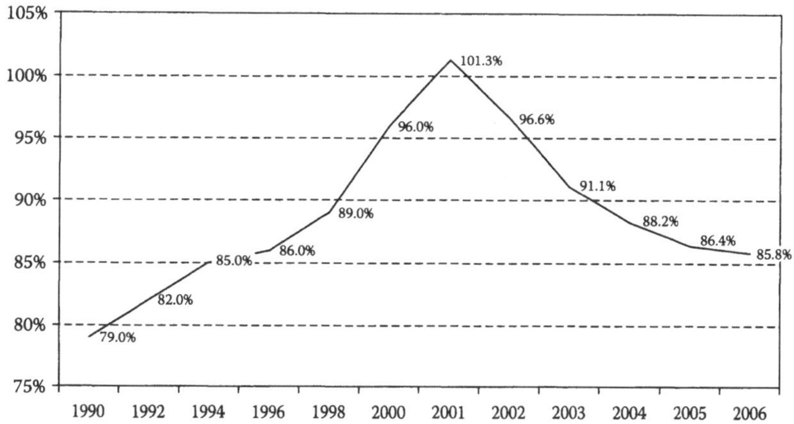

Figure 1.4 Aggregate funding ration for S&L public pension systems. (From National Association of State Retirement Administrators.)

This significant drop in funding ratio in recent years, however, is not unique to public pension plans. Private sector defined benefit plans also suffered a similar fate. For single-employer pension plans insured by the Pension Benefits Guaranty Corporation (PBGC), the overall funding ratio decreased from 144 percent in 2000 to 84 percent in 2003 (Pension Benefit Guaranty Corporation, 2006). Multiemployer pension plans insured by the PBGC saw their funding ratio decrease from 105 percent in 2000 to 64 percent in 2003. The total unfunded pension liability between the single and multiemployer pension plans stood at $440 billion, compared to total pension assets of $1.66 trillion. Partly because of this deterioration in pension funding ratio, Congress enacted a pension reform bill in 2006 to shore up the financial health of private defined benefit pension plans.

1.7 Overview

The rest of the book addresses various aspects of public pension plans management. Chapters 2 through 5 contain the four core elements in managing a defined benefit pension plan. Chapter 2 looks at the design issues of a defined benefit program and, most importantly, the issue of adequacy of pension benefits. Once the benefits are designed, Chapter 3 discusses how the future cost of such benefits is determined, how it is going to be financed, and how such financial information should be disclosed to the public. Chapter 4 looks at the fiduciary entity that manages pension plans with regard to such topics as governance structure, fiduciary responsibilities, financial reporting, funding policy, and oversight. Chapter 5 is devoted to one of the responsibilities of the fiduciary entity, namely investment management, due to its importance. It addresses investment theories and main practices that are derived from the theories when it comes to the management of pension asset investment.

In Chapter 6, the discussion is broadened to cast public pension plan management in the broader context of public finance. It examines the relationship between pension funding and government operating budgets and how they can affect each other. Several public pension plan management cases are discussed to illustrate this relationship and policy recommendations are suggested as how to manage this relationship for the betterment of both the pension plans and government operating budgets.

Chapter 7 further expands the discussion of public pension plans into considering the alternatives to defined benefits plans, such as defined contribution and hybrid plans. The discussion focuses on the advantages and disadvantages of various pension plan types.

Chapter 8 examines another retirement benefit, retirees' health benefit. This issue comes to the forefront of public finance due to the requirement of two accounting statements published in 2004 by the Governmental Accounting Standards Board. The discussion is focused on determining the cost of retirees' health benefits and dealing with such costs.

The final chapter (Chapter 9) examines broad policy issues on pension benefits in the public sector, with regard to its affordability and what can be done to ensure its sustainability in the future.

References

General Accounting Office. 1980. *An Actuarial and Economic Analysis of State and Local Government Pension Plans.* (PAD-80-1).

———. 1996. *Public Pensions: State and Local Government Contributions to Underfunded Plans*, (GAO/HEHS-96-56).

Hua, Thao. 2007. World's largest pension funds: $10 trillion strong. *Pensions & Investments*, September 3, 35(18).

Munnell, Alicia H. 2005. *Mandatory Social Security Coverage of State and Local Workers: A Perennial Hot Button.* http://crr.bc.edu/images/stories/Briefs/ib_32.pdf (Accessed June 3, 2007.)

Pennsylvania Public Employee Retirement Commission. 1992. *Special Report: Study of the Current Structure of Local Government Retirement Systems and Recommended Establishment of a Statewide Retirement System.* http://www.perc.state.pa.us/perc/lib/perc/1992_STATEWIDE_RETIREMENT_SYSTEM_FOR_LOCAL_GOVERNMENT_EMPLOYEES.pdf (Accessed May 31, 2007.)

Pension Benefit Guarantee Corporation. 12/21/2005. *An Analysis of Frozen Defined Benefit Plans.* http://pbgc.gov/docs/frozen_plans_1205.pdf (Accessed August 12, 2006.)

———. 2006. *The Pension Insurance Data Book 2005.*

Sterett, Susan M. 2003. *Public Pensions: Gender and Civic Service in the States, 1850-1937.* Ithaca, NY and London: Cornell University Press.

Social Security Administration. 2007. *Fact Sheet: Social Security.* http://www.ssa.gov/pressoffice/factsheets/basicfact-alt.pdf (Accessed September 19, 2007.)

U.S. Census Bureau. 1990. *1987 Census of Governments: Employee-Retirement Systems of State and Local Governments.* Washington D.C.: U.S. Government Printing Office.

———2002. *2002 Census of Governments: Government Organization.* Washington D.C.: U.S. Government Printing Office.———. 2004. *2002 Census of Government: Employee-Retirement Systems of State and Local Governments.* Washington D.C.: U.S. Government Printing Office.

U.S. Department of Labor, Bureau of Labor Statistics. 1981. *Employee Benefits in Industry, 1980.* Washington, D.C.: U.S. Government Printing Office.

———.1991. *Employee Benefits in Small Private Establishments,* 1990. Washington, D.C: U.S. Government Printing Office.

Appendix: Eighty-Three (83) State Pension Systems in the Wisconsin Survey

State	*State Code*	*Retirement System Name*	*System Acronym*
Alabama	AL	Employees' Retirement System	ERS
Alabama	AL	Teachers' Retirement System	TRS
Alaska	AK	Public Employees' Retirement System	PERS
Alaska	AK	Teachers' Retirement System	TRS
Arizona	AZ	State Retirement System	SRS
Arkansas	AR	Public Employees Retirement System	PERS
Arkansas	AR	Teacher Retirement System	TRS
California	CA	Public Employees' Retirement System	PERS
California	CA	State Teachers' Retirement System	TRS
Colorado	CO	Public Employees' Retirement Association	PERA
Connecticut	CT	State Employees Retirement System	SERS
Connecticut	CT	Teachers' Retirement System	TRS
Delaware	DE	Public Employees' Retirement System State Employees' Pension Plan	SEPP
Florida	FL	Florida Retirement System	FRS
Georgia	GA	Employees' Retirement System	ERS
Georgia	GA	Teachers Retirement System	TRS
Hawaii	HI	Employees' Retirement System	ERS
Idaho	ID	Public Employee Retirement System	PERS
Illinois	IL	State Employees' Retirement System	SERS
Illinois	IL	Teachers' Retirement System	TRS
Illinois	IL	Municipal Retirement Fund	MRF
Indiana	IN	Public Employees' Retirement Fund	PERF
Indiana	IN	Teachers' Retirement Fund	TRF
Iowa	IA	Public Employees' Retirement System	PERS
Kansas	KS	Public Employees Retirement System	PERS
Kentucky	KY	Employees Retirement System	ERS
Kentucky	KY	Teachers' Retirement System	TRS
Louisiana	LA	State Employees' Retirement System	SERS
Louisiana	LA	Teachers' Retirement System	TRSL
Maine	ME	State Retirement System	SRS

Appendix: Eighty-Three (83) State Pension Systems in the Wisconsin Survey (Continued)

State	*State Code*	*Retirement System Name*	*System Acronym*
Maryland	MD	State Retirement and Pension System	SRPS
Massachusetts	MA	State Employees' Retirement System	SERS
Massachusetts	MA	Teachers' Retirement System	TRS
Michigan	MI	State Employees' Retirement System	SERS
Michigan	MI	Municipal Employees' Retirement System of Michigan	MERS
Michigan	MI	Public School Employees Retirement System	PSERS
Minnesota	MN	Minnesota State Retirement System	MSRS
Minnesota	MN	Public Employees Retirement Association	PERA
Minnesota	MN	Teachers Retirement Association	TRA
Mississippi	MS	Public Employees' Retirement System	PERS
Missouri	MO	State Employees' Retirement System	SERS
Missouri	MO	Local Government Employees Retirement System	LAGERS
Missouri	MO	Public Schools Retirement System	PSRS
Montana	MT	Public Employees' Retirement Board	PERS
Montana	MT	Teachers' Retirement System	TRS
Nebraska	NE	Public Employees Retirement Systems	PERS
Nebraska	NE	School Retirement System	SRS
Nevada	NV	Public Employees' Retirement System	PERS
New Hampshire	NH	New Hampshire Retirement System	NHRS
New Jersey	NJ	Public Employees' Retirement System	PERS
New Jersey	NJ	Teachers' Pension and Annuity Fund	TPAF
New Mexico	NM	Public Employees Retirement Association	PERA
New Mexico	NM	Educational Retirement Board	ERA
New York	NY	State and Local Retirement System	ERS
New York	NY	Teachers' Retirement System	TRS
North Carolina	NC	Teachers' and State Employees' Retirement System	TSERS

Appendix: Eighty-Three (83) State Pension Systems in the Wisconsin Survey (Continued)

State	*State Code*	*Retirement System Name*	*System Acronym*
North Carolina	NC	Local Government Employees' Retirement System	LGERS
North Dakota	ND	Public Employees Retirement System	PERS
North Dakota	ND	Teachers' Retirement Fund	TRF
Ohio	OH	Public Employees Retirement System	PERS
Ohio	OH	State Teachers Retirement System	STRS
Oklahoma	OK	Public Employees Retirement System	PERS
Oklahoma	OK	Teachers' Retirement System	TRS
Oregon	OR	Public Employees Retirement System	PERS
Pennsylvania	PA	State Employees' Retirement System	SERS
Pennsylvania	PA	Public School Employees' Retirement System	PSERS
Rhode Island	RI	Employees' Retirement System	ERS
South Carolina	SC	South Carolina Retirement Systems	SCRS
South Dakota	SD	South Dakota Retirement System	SRS
Tennessee	TN	Consolidated Retirement System	CRS
Texas	TX	Employees Retirement System	ERS
Texas	TX	Teacher Retirement System of Texas	TRS
Texas	TX	Municipal Retirement System	MRS
Utah	UT	State Retirement Systems	SRS
Vermont	VT	State Employees' Retirement System	SRS
Vermont	VT	Teachers Retirement System	TRS
Virginia	VA	State Retirement System	SRS
Washington	WA	Public Employees' Retirement System	PERS
Washington	WA	Teachers' Retirement System	TRS
West Virginia	WV	Public Employees Retirement System	PERS
West Virginia	WV	Teachers' Retirement System	TRS
Wisconsin	WI	Wisconsin Retirement System	WRS
Wyoming	WY	Wyoming Retirement System	WRS

Chapter 2

Pension Benefit Design

To understand public pension plan management, the first step is to know what pension benefits are and how they are determined. This is important for at least two reasons. First, it will help us understand the cost of pension benefits and the financing of such costs, which is the subject of the following chapter. Second, it will help us understand the impact of any increase in pension benefits on the fiscal health of pension plans and the government's operating budget, which is the subject in Chapter 6.

In this chapter, we first examine in detail the various pension benefits offered in the public sector. Then, we briefly examine a broader issue of adequacy of pension benefits in the public sector. As discussed in the previous chapter, there are several different pension plans depending on the type of employment, each with its own set of pension benefit design. For the sake of simplicity and broader applicability, the discussion of pension benefit design is focused on pension plans that cover the two largest groups of employees, teachers and general government employees.

2.1 Regular Service-Related Benefits

Pension benefits can generally be divided into two main categories: benefits related to normal service and ancillary benefits. In this section, we discuss various benefits related to normal service, chiefly normal service benefits, early retirement benefits, postemployment benefit adjustment, and purchase of service credits.

A full-time employment status typically qualifies for participation in a public pension plan. However, because there are usually different plans for different categories of employees, such as general employees, uniformed officers, teachers, and judges, the type of work will determine in which particular plan the employee will participate.

2.1.1 Normal Service Benefit

Normal service benefit, or normal retirement benefit, accounts for the lion's share of all retirement benefits. It is called normal retirement benefit because the benefit can be received only when the employee reaches a normal retirement age. The normal retirement benefit is defined by a formula (leading to the term "defined benefit"):

$$\text{Final average salary} \times \text{Years of credited service} \times \text{Benefit multiplier}$$

Final average salary (FAS) is the average salary over the last few years prior to a member's retirement or termination of employment. Years of credited service are the number of years the retiree or the terminated employee has worked for one particular employer. Benefit multiplier (BM) means the percentage of final average salary the retiree can replace in her annual retirement benefit for each year of service. If the multiplier is 1.5 percent and the retiree has worked for the same employer for 30 years, then her annual retirement benefit will be equal to 45 percent of her FAS. While these three factors appear straightforward, there are many variations of these factors in actual pension benefit design. What follows is a detailed discussion of each of these three factors.

2.1.1.1 Benefit Multiplier

Of these three factors in the formula that determine the size of retirement benefit, the multiplier has the most impact. A relatively small increase in the multiplier can mean a significant increase in retirement benefit, as the increase applies to every year of service. For example, an increase in the multiplier from 1.5 percent to 1.65 percent means an increase of 10 percent in retirement benefit. While a constant multiplier for all years of service is the norm for most public pension plans, some plans also have a differential multiplier structure, with increasingly higher multipliers applied to longer years of service.* By applying higher multiplier to later years of service in one's career, it encourages an employee to work longer and also to stay with the same employer. Another practice that is fairly common in the public sector is that there are significant differences in multipliers among different classes of employees. The most notable example is that public safety employees tend to have higher multipliers than other classes of employees as the former have a shorter working career due to the hazardous and demanding nature of their employment and, thus, need a higher multiplier to have a comparable level of pension benefit in terms of the percentage of preretirement salary being replaced.

---

* For example, for the first 10 years, the multiplier is 1.5 percent and for the next 10 years the multiplier increases to 1.6 percent and, for years beyond that, the multiplier goes up to 1.7 percent.

Even within a similar class of employees, there can still be significant variation in the multiplier from one state plan to another. Such variation can be accounted for by at least two important factors. The first is the status of Social Security participation. As mentioned in the first chapter, about a quarter of public sector employees do not participate in the Social Security program. Everything else being equal, the multiplier should be higher for a pension plan that does not participate in the Social Security program than for one that does. It is also easier for the pension plan not in the Social Security system to provide a higher multiplier and, thus, higher pension benefits as the part of Social Security taxes that the employer would otherwise pay to the federal government can be used to boost the pension benefit.

The second major factor that affects the multiplier is whether a pension system is contributory or noncontributory. A noncontributory system means that the employees are not required to make any contribution toward their retirement benefits. There are nine pension systems that are noncontributory (see Appendix, Chapter 1 for acronyms): Arkansas's PERS and TRS, Connecticut's SERS, Florida's FRS, Hawaii's ERS, Michigan's SERS, Missouri's SERS, Tennessee's CRS, and Utah's SRS (Wisconsin Legislative Council, 2005).* Everything being equal, the multiplier should be somewhat lower for noncontributory systems than for contributory systems, as the combined employer and employee contribution rate should be lower for the noncontributory system. Even for those contributory systems, there can still be significant variation in employee contribution rate. For example, the member contribution rate was just about 1 percent for Georgia ERS, whereas it was 10 percent for Missouri PERS in 2004. For the same reason, a higher employee contribution rate generally should also be correlated with a larger multiplier, holding the employer contribution rate constant.

Based on the survey of 83 state-level pension systems by the Wisconsin Legislative Council, the pension systems are divided into eight groups depending on the level of multiplier and Social Security coverage. The multiplier is divided into four groups: under 2 percent, 2.0 to 2.4 percent, 2.5 percent and over, and those with ranges that span multiple categories. The results are shown in Table 2.1. Two observations can be made about the benefit multiplier from this table:

1. While there is substantial difference in the level of multiplier, from just over 1 percent to 3 percent, the vast majority of systems cluster in a range from 1.5 to 2.5 percent, with 2 percent used most frequently and many other multipliers close to 2 percent.
2. As a group, the systems not covered by Social Security have a higher multiplier than those covered by Social Security. There are no systems in the

* Unless otherwise noted, the data on the characteristics of state-level pension systems are drawn from the findings of the Wisconsin Legislative Council's survey of state pension systems in 2004.

Table 2.1 Normal Retirement Benefit Multiplier

	Social Security Coverage	*Non-Social Security Coverage*
1.0–1.99% (31 systems)	AR PERS (1.72%), AR TRS (1.39%), CT SERS (1.33–1.625%), DE SEPP (1.85%), FL FRS (1.6%), HI ERS (1.25%), IL SERS (1.67%), IN PERF (1.1%), IN TRF(1.1%), KS PERS (1.75%), KY ERS (1.97%), MD SRS (1.8%), MI SERS (1.5%), MI PSERS (1.5%), MN MSRS (1.7%), MN PERA (1.7%), MN TRA (1.7%), MO SERS (1.7%), NH NHRS (1.515–1.67%), NJ PERS (1.82%), NJ TPHF (1.82%), NC TSERS (1.82%), NC LGERS (1.85%), OR PERS (1.5%), SC SCRS (1.82%), SD SRS (1.55–1.625%), TN CRS (1.5%), VT SRS (1.67%), VT TRS (1.67%), VA SRS (1.7%), WI WRS (1.6%)	
2.0–2.49% (26 systems)	AL ERS (2.0125%), AL TRS (2.0125%), AZ SRS (2.1–2.3%), GA ERS (2%), GA TRS (2%), ID PERS (2%), MT TRS (2%), NE SPP (2%), NM ERA (2.35%), NY TRS (2%), ND PERS (2%), ND TRF (2%), OK PERS (2%), OK TRS (2%), TX ERS (2.3%), UT SRS (2%), WA PERS (2%), WA TRS (2%), WV PERS (2%), WV TRS (2%), WY WRS (2.125-2.25%)	CA TRS (2–2.4%), CT TRS (2%), IL TRS (2.2%), ME SRS (2%), TX TRS (2.3%)
2.5% and over (9 systems)	NM PERS (3%), PA SERS (2.5%), PA PSERS (2.5%)	CO PERA (2.5%), KY TRS (2.5%), LA SERS (2.5%), LA TRSL (2.5%), MO PSRS (2.5–2.55%), NV PERS (2.67%)
Others (15 systems)	CA PERS (2–2.5%), IL MRF (1.67–2%), IA PERS (1–2%), MI MERS (1.3–2.5%), MS PERS (2–2.5%), MO LAGERS (1–2%), MT PERS (1.785–2%), NY ERS (1.67–3.5%), RI ERS (1.7–3%)	AK PERS (2–2.5%), AK TRS (2–2.5%), MA SERS (.5–2.5%), MA TRS (.1–2.5%), OH PERS (2.2–2.5%), OH STRS (2.2–2.5%)

Source: Wisconsin Legislative Council 2004 Comparative Study of Major Public Employee Retirement Systems.

Table 2.2 Limitation on Pension Benefits

	State Funds
50–94.5% of FAS (16 systems)	CT TRS, IL SERS, IL TRS, IL MRF, IA PERS, MA SERS, MA TRS, MI MERS, NV PERS, NM PERS, RI ERS, TN CRS, VT SRS, VT TRS, WI WRS
100% FAS (15 systems)	CA TRS, CO PERA, FL FRS, ID PERS, KY TRS, LA SERS, LA TRSL, MD SRS, MN PERA, MN TRA, MS PERS, MO PSRS, OH PERS, OH STRS, TX ERS, VA SRS

Source: Wisconsin Legislative Council 2004 Comparative Study of Major Public Employee Retirement Systems.

former group with a multiplier below 2 percent and most of them have a multiplier around 2.5 percent.

While not directly observable from this table, there is some correlation between the employee contribution rate and the benefit multiplier. For example, of the nine noncontributory systems, the multipliers of eight systems are well below 2 percent and only Utah has a multiplier of 2 percent. Of those systems with a multiplier over 2.5 percent, New Mexico has a contribution rate of 7.6 percent; Pennsylvania SERS and PSERS, 6.25 and 7.5 percent, respectively; and Nevada PERS 10.5 percent, all above the average employee contribution rate of 5 percent.

Because of the difference in the Social Security coverage and employee pension contribution rates, any comparison in pension benefit level between pension systems has to take these two factors into account.

It is conceivable that with sufficient years of service, final pension benefits can exceed final average salary. For example, 40 years of service with a multiplier of 2.5 percent can lead to 100 percent of final average salary. Some public plans set a cap on pension benefit in terms of the percentage of final average salary. Table 2.2 shows the limit put on the final benefit, based on the Wisconsin survey in 2004.

For those that do not set a limit, the vast majority have a multiplier that is at or below 2 percent, making it unlikely for the pension benefit to exceed 100 percent of final average salary.

2.1.1.2 Final Average Salary

The second important factor in determining the size of benefit is final average salary (FAS). The most important design issue here is the number of years used to determine the FAS. Since salary typically goes up every year, the fewer the number of years needed to determine the FAS, the greater the FAS and, thus, the greater the annual pension benefit. In the extreme case, the FAS can simply be the last year's salary. The purpose of using FAS rather than the actual final year's salary is to mitigate the effect of spiking in the final year's salary on the final pension benefit.

Table 2.3 Period for FAS Determination

Number of years in the period	1	2	3	4	5
Number of systems	2	2	54	6	17

Source: Wisconsin Legislative Council 2004 Comparative Study of Major Public Employee Retirement Systems.

If the FAS is simply the final year's salary, then a large increase in the final year's salary, say due to overtime or a large salary increase, can substantially inflate the final retirement benefit and also makes it more difficult to predict beforehand a member's lifetime retirement benefits. By using the average of salaries over several years, then the effect of spike in the final year will become more muted. To further prevent the effect of excessive salary growth in the final year on the eventual level of pension benefit, plan sponsors sometimes also adopt a policy of limiting the salary growth to a certain percentage in the final year. Any increase above that rate will not be used in calculating FAS.

One variation of the FAS calculation is to pick three or five highest annual salaries over a longer period of time, say the last 10 years before the retirement, rather than just the salaries for the final three to five consecutive years. This is done to recognize that some employees may earn higher salaries earlier in the period and this variation can maximize their pension benefits. According to the Wisconsin survey in 2004, 10 pension systems adopted this variation in calculating the FAS, typically using the three highest annual salaries over the last 10 years. Table 2.3 shows the number of years used by public pension systems in calculating FAS based on the Wisconsin survey.*

Clearly, three years is by far the most common period over which the final average salary is calculated in the public sector. The two systems that used the final year's salary as the FAS in 2004 were California's PERS and TRS. Since the survey, these two systems switched to using the final three years in determining the FAS.

2.1.1.3 Years of Service

While it seems fairly straightforward to determine the value of the third factor, years of service, in calculating the retirement benefit, there are at least three issues related to years of service that render this factor more complicated in determining the retirement benefit. The first issue is the vesting requirement. Simply participating in a pension plan does not make one eligible for pension benefits. To be eligible for normal and other pension benefits discussed later, the participant also has to meet the vesting requirement, requiring an employee to work for a number of years before she is entitled to (or vested in) any pension benefits. If she terminates her

* This table does not make a distinction between systems using the final three years or the highest three years over a 10-year period.

Table 2.4 Vesting Requirement

Number of years	0	2	3	4	5	6	8	10
Number of systems	2	1	6	4	46	2	4	18

Source: Wisconsin Legislative Council 2004 Comparative Study of Major Public Employee Retirement Systems.

employment before she is vested, then she forfeits the pension benefits accrued. Due to turnover, one purpose of vesting requirement is to avoid unnecessary administrative burden to keep track of the benefits of employees who terminate employment after only a short period of time. Other than this main purpose, the vesting requirement also serves another purpose. The vesting requirement provides an incentive for the employee to stay with one employer. The longer the vesting period, the more costly it becomes for the employee to terminate employment before vesting and, thus, the employee will more likely stay. Table 2.4 shows the vesting requirement for public pension plans based on the Wisconsin survey in 2004.

By 2004, the five-year vesting period has become the most commonly used vesting requirement among public pension plans. An important trend in vesting requirement, which is not observable from this table, is that the average vesting period has been dramatically reduced over time. In 1982, more than half of the pension systems in the Wisconsin survey required 10 years for vesting. By 2004, only 18 pension systems still used a 10-year vesting requirement. This makes it easier for employees to be vested for pension benefits and, thus, makes the benefits more portable. Zero years in the table means immediate vesting. The two states with immediate vesting are Arizona and Wisconsin.

The second issue has to do with whether these years of service occur with one employer or multiple employers. As the retirement benefits are tied to FAS, the accrual of retirement benefits is heavily backloaded rather than evenly spread out over the retiree's working career.* Such backloading of benefit accrual rewards a retiree who works for the same employer for his or her entire working career. For two retirees with the same FAS, the retiree who does not switch jobs will earn greater retirement benefits compared to a retiree who has switched jobs several times in her career because the latter's highest FAS before her retirement applies to fewer years than the former.† Therefore, such formula-based benefit accrual is said

* If the accrual of pension benefit grows at a fairly constant rate, then it is evenly spread out over one's working career. In a defined benefit pension plan, the accrual of pension benefit grows at a much faster rate in later years than in early years, due to the interaction between final average salary and years of service, and, thus, it is considered backloaded.

† A quick example makes this clear. Suppose Employee A works for Employer I for 15 years with a FAS of $40,000 and Employer II for another 15 years with a FAS of $60,000. Employee B works for Employer I for the entire 30 years with a FAS of $60,000. If both pension plans use a multiplier of 2 percent, then Employee A's final pension is $30,000, whereas Employee B's final pension is $36,000, 20 percent greater.

to provide an incentive to the employees to stay with one employer. It also makes the benefit accrual less portable from one employer to another. To mitigate this effect, public pension plans allow for purchase of service credit, which is discussed later in this chapter.

The third issue is the conversion of unused sick and/or annual leave to years of service. At the end of the career, many employees have accumulated unused sick and/or annual leave. Most public pension plans allow employees to convert such unused leave to years of service credit. Such conversion provides two benefits to the employees. First, it increases the years of service and, thus, the pension benefit. Second, it also allows employees to retire earlier and still qualify for normal pension benefits. However, as like most other characteristics, there is significant variation among public pension plans in terms of conversion. For example, Oklahoma's PERS allows a maximum of half a year of service credit that can be converted from unused sick leave, whereas Illinois state pension plans allows up to two years of service credits that can be converted from unused sick leave. The conversion rate can also be quite different.

2.1.1.4 Requirement for Normal Retirement Benefits

In order to receive the normal retirement benefits, the retiree also has to meet the age requirement for such benefits. This is the age at which an employee can retire and receive unreduced pension benefit determined by the formula mentioned above. This is similar to the normal age requirement for receiving unreduced Social Security benefit. While there is only one normal retirement age for Social Security benefit, which is 65 for people born in 1937 or earlier, gradually increasing to 67 for people born in 1960 and later, there are three primary ways of determining the requirement for receiving unreduced normal retirement benefits in the public sector: age, years of service, or a combination of both.

1. **Age**: If the requirement for normal retirement is based on age, then the employee has to reach a certain age, say, 60 years old. Such age requirement is usually accompanied by a requirement for minimum years of service, say, five years, which is similar in nature to a vesting requirement. Sometimes a public plan can have more than one normal retirement age, with normal retirement at a younger age accompanied by a longer service requirement.
2. **Years of service**: Under this requirement, an employee can retire at any age and collect normal retirement benefit as long as she has worked for a certain number of years. Such years of service requirement can range anywhere from 20 to 35 years.
3. **Combination of age and years of service**: Under this requirement, which is a combination of the previous two requirements, an employee can retire with normal benefit as long as her age and years of service add up to a certain number. For example, if an employee's age and years of service add up to 80, then she will be eligible for normal retirement benefits. This is known as the "Rule

Table 2.5 Requirement for Normal Retirement Benefits

Requirement	Any 25	Any 30	Any 35	Rule 80	Rule 85	Rule 90	60/5	60/10	65/5 or less
Systems	5	18	4	7	9	6	16	9	26

Source: Wisconsin Legislative Council 2004 Comparative Study of Major Public Employee Retirement Systems.

of 80." This type of rule provides more flexibility for employees in meeting the requirement for normal pension benefits.

For many public pension plans, members can use more than one requirement to determine normal retirement. For example, Missouri's PERS allows members to use any one of the three requirements mentioned above to determine normal retirement. With the combination of different ages and different years of service, there can be a plethora of requirements for normal benefits. As it is unlikely to list all the requirements used by public pension systems, Table 2.5 lists only some of the more frequently used requirements, which are based on the 2004 Wisconsin survey.

"Any 25" means that an employee can collect normal retirement benefit at any age after 25 years of service, and "60/5" means that the employee can retire at age 60 with 5 years of service. This table illustrates a wide range of what is considered a normal retirement age in the public sector. For example, the "Any 25" requirement means that an employee can potentially retire at age 50 with normal pension benefit.

2.1.1.5 State Income Tax Treatment

One further complication in comparing the pension benefits is state income tax treatment. There are three income tax treatments. First, pension benefits are not taxed, either because there is no state income tax in that state or the state government specifically exempts such benefits from taxation. Second, pension benefits are exempt to a certain point and the amount above that is taxable. The partial exemption amount varies significantly, from only $2000 in West Virginia to $20,000 in Colorado. Third, all pension benefits are taxable.

Tax exemption is particularly valuable for states with high income tax rates. For example, the public pension benefit is exempt from state income tax in New York State, which has a top income tax rate of 6.85 percent. Table 2.6 shows the tax treatment of pension benefits.

2.1.2 Early Retirement Benefit

In addition to normal retirement benefits, public pension plans also allow for early retirement benefits, which is offered to those who retire before they meet the requirement for normal benefit. To qualify for early retirement benefits, the employee also

Table 2.6 State Income Tax Treatment

State Income Tax Treatment	*Exempt*	*Partial Exempt*	*No Income Tax*	*Taxable*
Number of systems	20	31	12	20

Source: Wisconsin Legislative Council 2004 Comparative Study of Major Public Employee Retirement Systems.

Table 2.7 Requirement for Early Retirement Benefits

Requirement	*55/5 or less*	*55/10–20*	*Any 25*	*None*
Systems	22	16	15	8

Source: Wisconsin Legislative Council 2004 Comparative Study of Major Public Employee Retirement Systems.

has to meet minimum age and/or years of service requirement, which is very similar in nature to that for normal retirement. The only difference is that the age and years of service limits are all lowered for early retirement benefits. Just as there are many variations of normal retirement requirement, there are also many ways of setting early retirement requirement. Table 2.7 lists a few of the commonly used requirements, based on the Wisconsin survey in 2004. Eight pension systems do not have early retirement benefits. These are the systems that tend to have a relatively low age and years of service threshold for normal retirement benefits to begin with.

While the defined benefit formula described above for determining normal retirement benefit is also used for determining early retirement benefit, pension plan sponsors add an adjustment (or reduction factor) to the early retirement benefit determined by such a formula. As early retirement lengthens the period over which the retiree will draw on benefits and shortens the period over which pension contributions are made, it increases the cost to the pension plan. To make early retirement "cost neutral," pension plans, therefore, reduce by a certain percentage the early retirement benefit calculated using the formula discussed above. This percentage is usually applied to each year short of the requirement for normal retirement. While some plans apply a constant percentage to each and every year, others use differential percentages. For example, in the case of a constant percentage of 5 percent a year, if the normal retirement age is 60 and an employee retires at the age of 55, then her retirement benefit will be reduced by 25 percent. In case of a differential percentage structure, the Oklahoma PERS has a month-by-month reduction rate between age 55 and 62. At age 55, the retiree can get 60 percent of her accrued benefit. At age 56, it increases to 63.3 percent. This reduction of 3.3 percent from age 56 to 55 is smaller than the reduction from 62 to 61, which is 6.7 percent (Oklahoma Public Employees Retirement System, 2006). Table 2.8 shows

Table 2.8 Early Benefit Reduction Factor

Reduction Factor	*3 percent*	*5 percent*	*6 percent*	*Table*
Systems	14	5	21	18

Source: Wisconsin Legislative Council 2004 Comparative Study of Major Public Employee Retirement Systems.

some of the most frequently used reduction percentages for early retirement benefit, based on the Wisconsin survey in 2004.

2.1.2.1 Early Retirement Incentive Program

While the early retirement discussed so far is initiated by the employee herself, there are also early retirement incentive programs (ERIP) initiated by government employers from time to time. Such a program is designed to induce employees to elect early retirement on favorable terms. In return for retiring prior to normal retirement age, employees receive incentives that may include decreased early retirement reduction factors, enhanced service credits, or even unreduced retirement benefits. Employers often use ERIP for permanent workforce reduction and/or restructuring, or to achieve short-term payroll savings for budgetary relief. The success of an ERIP depends on the comparison between savings and cost. The savings can be achieved in two ways. First, if some of the early retirees are not replaced, this leads to a permanent reduction in workforce and results in both short-term and permanent employment cost savings. Second, if some of the vacant positions are filled with employees at a lower salary level, it also leads to cost savings. Such savings will be offset by increased pension costs. The cost of ERIP results from the increase in pension liabilities due to the fact that early retirees will receive pension benefits (unreduced in many cases) for a longer period of time and the pension plan will also receive fewer years of contribution from both the employer and the employee.

As the incentives will entice some employees to retire early, the savings in personnel cost are immediate, and achieving short-term savings for budgetary relief is almost guaranteed through an ERIP since the full cost of ERIP will not materialize for many years to come. To achieve long-term cost savings, however, is far more difficult, as either a significant portion of the early retirees' positions cannot be filled, or such positions have to be filled with employees at a much lower salary and benefit level. In a study on the financial impact of a Pennsylvania state government ERIP in the early 1990s, the Pennsylvania Public Employee Retirement Commission (1995) found that at the breakeven point where the savings would exactly equal the cost, the replacement rate would have to be 69 percent. This makes it difficult to assess the long-term financial impact of an ERIP in the public sector because, in the long run, it is difficult, if not impossible, for the government to control to what extent these vacant positions will remain unfilled. This is unless there is legislation accompanying such ERIP mandating that some positions are permanently eliminated. Because of this inherent difficulty in the public sector, most of the ERIPs in the public sector were unsuccessful, based on the study of Pennsylvania Public Employee Retirement Commission (1995). An analysis of a more recent ERIP in New Jersey also shows that the long-term cost is more than the savings (Craven-McGinty and Chen, *New York Times*, June 15, 2007).

2.1.3 Postemployment Benefit Adjustment

For employees, it is important to know not only what the pension benefit is at the time of retirement, but also whether such pension benefits will be adjusted for cost of living in the future. Without such adjustment, inflation will gradually erode the purchasing power of pension benefit over time. Because an average retiree is expected to live for many years after retirement, such adjustment is a significant benefit related to the normal or early retirement benefit for the employee. At the same time, it also adds significant cost to the pension plan. In pension benefit design, perhaps the most basic dilemma concerning postretirement adjustments is whether to implement them on an ad hoc or automatic basis. The choice involves a trade-off between flexibility for the employer and financial security for the retirees.

Ad hoc adjustments are flexible in the sense that they can be designed in any way the plan sponsors desire. They can grant a one-time benefit increase in a set amount for every retiree or a set percentage of her pension benefit. The biggest advantage of ad hoc adjustment to the plan sponsor is that it is not locked into a long-term commitment of annual automatic increase of pension benefits. This is particularly beneficial when the pension plan is either temporarily underfunded, due to a sharp drop in the value of financial assets, or chronically underfunded. The lack of an automatic benefit adjustment will reduce the growth of pension liability and allow for more assets to stay in the plan for investment and better chance for improvement in pension funding ratio. The disadvantage of ad hoc adjustment is that the retirees do not have the assurance that there will be adjustment year after year even though annual inflation rates rarely turn negative. Another disadvantage is that it may be more difficult for system actuaries to project the growth of future pension liability and, thus, pension contribution, as the timing and magnitude of future adjustments cannot be determined. That means ad hoc adjustments cannot be prefunded when the employees are still working.

There are generally two different types of ad hoc adjustment. One is a true ad hoc adjustment, which gives the plan sponsors total discretion over adjustment with no conditions set as to when the adjustment needs to be made. The second type of ad hoc adjustment has conditions attached as to when the adjustment should be made. Such a condition is usually tied to investment performance. For example, in Wisconsin, annuities are increased annually if the investment income credited to retired life funds is in excess of the assumed rate of return and the resulting adjustment would be at least 0.5 percent (Wisconsin Retirement System, 2005).

An automatic benefit adjustment, by definition, means that such an annual adjustment is required by a statute without the plan sponsor taking any specific action. The difference between plans with such automatic adjustment is whether the rate of adjustment is fixed or variable. The fixed rate method is simply to set a constant percentage increase every year. For example, the Florida Retirement System increases the pension benefit by 3 percent every year. The variable rate method links the adjustment to a certain inflation index. Since the Consumer Price Index (CPI) is the most widely used inflation index, the automatic adjustment is usually tied to the CPI. The variable rate can be anywhere from 50 to 100 percent of CPI. One advantage this method has over the fixed rate method is that it assures a certain relationship between the adjustment and the actual need. The disadvantage of a variable rate adjustment tied to CPI is that potentially it can be very expensive in an inflationary environment. To limit the plan's pension liability due to a large increase in the underlying inflation rate, the plan usually sets a cap on the size of adjustment. For example, the formula for pension benefit adjustment for Louisiana Teachers Retirement System is 100 percent of CPI with a 4 percent cap.

The advantage of automatic adjustment to the retirees is obvious. The retirees have more financial security as their retirement benefits will keep up with inflation, even if not to the fullest extent in some circumstances when the adjustment is not fully indexed to the inflation. The disadvantage of automatic adjustment to the plan sponsor is also obvious, as it makes it inflexible for the public employer, especially when the pension plan is underfunded. Such inflexibility is offset by the predictability of the pension benefit increase and better financial planning. Since the cost of benefit increase can be estimated, it can be advance funded and spread over the working life of the employee, making the cost just part of the regular annual pension contribution.

Based on the survey result of 83 state level pension systems by Wisconsin, they are put into five groups depending on how postretirement pension benefit adjustment is determined: ad hoc, excess investment earning, a constant percentage increase, 100 percent of CPI with a cap, or partial CPI with or without a cap. The results are shown in Table 2.9.

Most of the pension systems have automatic adjustment, and annual increase of 2 to 3 percent is fairly typical for both plans with a fixed rate and plans that cap the adjustment tied to the CPI.

2.1.4 Purchase of Service Credits

An important pension right of public pension plan members is the purchase of service credits. Because of the portability issue with defined pension benefits, which can lead to loss of or reduction in pension benefits, government employers allow

Table 2.9 Postemployment Pension Benefit Adjustment

Ad Hoc	AL ERS, AL TRS, DE SEPP, IN PERF, IN TRF, KS PERS, KY TRS, LA SERS, NH NHRS, NC TSERS, NC LGERS, ND PERS, ND TRF, OK TRS, PA SERS, PA PSERS, TX ERS, WV PERS, WV TRS, TX TRS
Excess earning	AZ SRS, CT TRS, IA PERS, WI WRS
Fixed	AR PERS (3%), AR TRS (3%), CA PERS (2%), CA TRS (2%), CO PERA (3.5%), FL FRS (3%), HI ERS (2.5%), IL SERS (3%), IL MRF (3%), IL TRS (3%), MI SERS (3% $300 annual cap), MI PSERS (3%), MS PERS (3%), MT PERS (3%), MT TRS (1.5%), NM PERA (3%), NV PERS (2%-5%), OK PERS (2.5%-4.5%), RI ERS (3%), SD SRS (3.1%)
100 Percent of CPI with a cap	GA ERS (1.5%), GA TRS (1.5%), ID PERS (1%–6%), LA TRSL (3%), KY ERS (5%), MA SERS (3%), MA TRS (3%), MD SRS (3%), ME SRS (4%), MN MSRS (2.5%), MN PERA (2.5%), MN TRA (2.5%), MO LAGERS (4%), NE SPP (2.5%), OH PERS (3%), OH STRS (3%), OR PERS (2%), SC SCRS (4%), TN CRS (3%),UT SRS (4%), VA SRS (5%), WA PERS (3%), WA TRS (3%), WY WRS (3%)
Partial CPI with or without cap	AK PERS (50%–75%), AK TRS (50%–75%), CT SERS (60%–75%), MO PSRS (80%), MO SERS (80%), NJ PERS (60%), NJ TPAF (60%), NM ERA (50%), NY ERS (50%), NY TRS (50%), TX MRS (70%), VT SRS (50%), VT TRS (50%)

Source: Wisconsin Legislative Council 2004 Comparative Study of Major Public Employee Retirement Systems.

pension plan members to purchase service credits in order to enhance their future pension benefits.

There are three main reasons for the "portability loss" of pension benefits. The first is the vesting requirement, as discussed earlier. An employee loses the right to accrued benefits if she terminates employment before becoming fully vested in a plan. Second, as the accrual of pension benefits is backloaded, an employee who changes job several times in her working career will have a lower total pension benefit at the time of retirement compared to someone else who stays in one job for her entire career. Third, for those plans that base the normal retirement age on a combination of age and years of service, an employee who stays in one job can retire at an earlier age with unreduced benefits compared to someone who switches job more than once in her career. If the "job switcher" wants to retire at the same age, it will be considered an early retirement and, thus, the retirement benefit will be reduced.

Purchase of service credits allows an employee to purchase credits for years of service not with the current employer. This will increase the years of creditable service with the final employer so that the final higher salary can be applied to more years of service. Also for some employees, purchasing service credit for several years will allow them to retire at an earlier age with unreduced pension benefit. In this case, the benefit of the purchase is equal to the early retirement discount. The purchase also allows the employee to qualify for early retirement at an earlier age if she so chooses.

There are four major design issues when it comes to purchase of service credits: (1) types of service allowed, (2) years of service allowed to be purchased, (3) the cost of purchase, and (4) the payment method.

2.1.4.1 Types of Service and Years of Service to Be Purchased

The types of prior service allowed for purchase are limited to those that are public and personal in nature. Services that are public in nature refer to those performed in the public sector with federal, state, and/or local government agencies as well as military service. Public pension systems also allow members, who terminate membership and later return to service with the same public employer, to purchase back years of service prior to termination. Services that are personal in nature refer to the leave of absence an employee takes to further education or give birth to and raise a child.

The number of years of prior service that can be purchased varies from one plan to the next. While it is fairly common among many plans to allow five years of prior service to be purchased, some plans allow for more years. For example, New Jersey allows for as many as 10 years of prior service to be purchased. To be eligible for purchasing service credit, the employee also needs to be vested and to become eligible for pension benefit. Another requirement is that the employee cannot be eligible for pension benefits for the same service with her previous employer for which she will receive pension benefit.

2.1.4.2 Cost of Purchase and Its Payment

Public plans vary in the percentage of the total cost the employee is required to pay for the service credits. Most plans that allow the purchase of prior service require the employee to pay at least the total amount she would have been required to contribute if service had been earned under the new system, and some systems may require payment of the full actuarial cost of the benefits purchased. The actuarial cost represents the total cost in today's dollars to pay for the increased benefits the employee will receive over her lifetime due to the purchase of service credits. This calculation will be based on many factors, such as the employee's current age and her assumed retirement age, her current salary and her final assumed salary, number of years purchased, and interest rate.* Because the cost is calculated at present value, the closer to retirement age when the purchase is made, the more costly it becomes.

* A full explanation of the actuarial cost valuation is provided in Chapter 3.

The cost that is the easiest to calculate is one for the purchase of previous service credits with the same employer. When an employee terminates her employment, she will get a refund of all her contributions. In this case, all she needs to do is pay back the refund to the pension plan, plus interest on the refund for the period between her termination and the date of purchase. The interest part is needed because her previous contributions, had they not been refunded, would have been invested to earn income. For the purchase of credit for leave of absence, first, the salary base needs to be determined. The required member payment would then be determined by applying the applicable member contribution rate to the compensation base. This amount, plus the interest until the date of payment, is the total cost to the employee. Similarly, when an employee purchases credits for prior service with another employer, she will also have to pay at least her share of the increase in her pension cost due to the purchase, determined by the estimated salary base and member contribution rate, plus interest. For some pension plans, a distinction is made between the types of service for which the employer will pay for all or part of the purchase cost. For example, with New Jersey's PERS, the employee has to pay for full cost of purchase of credit for military service, U.S. government civilian service, and local retirement system service, whereas both the employee and employer will share the cost of purchase of credit for all other services, including leave of absence.

To make it easier for the employee to figure out the cost of purchase, some pension plans also publish a purchase rate schedule, like the one shown in Table 2.10, which lists select purchase rates for different ages. The purchase cost will simply be the product of the purchase factor, annual salary, and number of years purchased.

There are two principal ways for the employee to pay for the purchase: lump-sum payment or payroll deduction. Once the total cost of the purchase is calculated by the plan, the employee can pay the cost in one lump sum, possibly with funds rolled over from her previous pension plan. More frequently, the total sum can also be amortized over a number of years through payroll deduction. While this makes it financially more feasible to purchase the service credit, it also cost more over time because the employee has to pay interest on the amount still owed to the pension plan.

Table 2.10 Purchase Rate

Age	*Purchase Factor*	*Age*	*Purchase Factor*
35	0.039469	50	0.055163
40	0.0436313	55	0.063145
45	0.048761	60	0.073142

Source: New Jersey Public Employees' Retirement System (2005).

2.2 Ancillary Benefits

In addition to the main retirement benefits mentioned above that account for the bulk of pension benefits, there are some other ancillary benefits offered by a defined benefit plan, chief among which are disability, death, health, and withdrawal benefits.

2.2.1 Disability Benefits

To claim disability benefits, one must be permanently disabled and no longer able to work. The calculation of a disability benefit is fairly similar to that of regular retirement benefits, which is the product of final average salary (FAS), years of service, and a benefit multiplier. In the disability benefit case, the FAS is the average salary prior to her disability rather than to her retirement. However, since disability is unpredictable and may occur early in one's career, the disability benefit based on such a formula can be very small. To provide a minimum living standard for permanently disabled employees, a pension plan typically provides a minimum disability benefit. The disabled employee will choose the greater of the minimum guarantee and the disability benefit based on the formula.

Just like the vesting requirement for regular benefits, employees are also required to work for a minimum number of years before they are eligible for disability benefits. One purpose of such minimum years of service is to prevent someone from seeking employment for the sole purpose of getting disability benefits if she knows she may be disabled in the near future. However, most pension plans make a distinction between whether the disability is work related or not. If the disability is work related, such vesting requirement is either reduced or eliminated. If work related, the employee will also get workers' compensation, which will reduce the disability benefit. The total benefits for work-related disability, including workers' compensation, are also higher than those for disability not related to work.

Using Virginia as an example, if the injury is not work related, the amount is the greater of the formula amount, or the minimum guarantee amount, which is set to one-third of the employee's FAS if she qualifies for full Social Security benefits, or 50 percent of her FAS if she does not qualify for full Social Security benefits. If the employee is injured on the job, then her disability benefits will be reduced by the amount of workers' compensation she receives. If the workers' compensation is equal to or greater than the disability benefits, the employee will receive no disability benefits. The minimum guaranteed benefit in work-related injury is also higher, 50 percent of the FAS if qualifying for primary Social Security benefits and two-thirds of FAS if not qualifying for primary Social Security benefits.

The disability benefit is also different for different classes of employees, such as regular employees versus police officers, with the latter receiving higher disability benefit. For example, in Florida, a member of the regular class will receive a disability benefit, which is the greater of 42 percent of FAS or actual earned benefit based on years of creditable service. A member of the Special Risk Class or Special Risk Administrative Support Class will receive a disability benefit equal to the greater of 65 percent of her final average salary or the actual earned benefit based on her years of creditable service.

In some pension systems, such as State Employees' Retirement System of Illinois (2005), the disability benefit ends when the disabled person reaches normal retirement age and the disability benefits will be replaced by the normal retirement benefits. While she receives disability benefits, her retirement account will continue to be credited with service and contributions as if she were working.

To continue receiving disability benefits, the disabled person also needs to get periodic medical examinations to determine the disability status. If she is deemed fit for work, then the disability benefits will stop. Because of the initial and continuing determination of disability status on the part of the pension plan administration, which is inherently more difficult than the determination of eligibility of most other benefits, it can lead to an administrative hearing and adjudication in some cases.

2.2.2 Death Benefits

Death benefits are also called survivor benefits, as it is a benefit to the survivors of the deceased member of a pension plan. There are two types of death benefits, one for those who die while still in service and one for those who die in retirement. If it is a death in service situation, the death benefit also depends on whether the active member is vested or not. If the member is not vested, the designated beneficiary will receive a refund of the member's contribution plus interest. If the member is already vested, the survivor will receive a monthly payment or a lump-sum equal to the present value of the monthly benefit. The monthly death benefit the survivor will receive is based on the benefit the deceased employee has accrued as of the date of death. The survivor benefit will be payable for the remainder of the survivor's life. While the survivor primarily refers to the spouse of the deceased member, it can also refer to her surviving parents or children. In the case of the spouse, the monthly death benefit can be withdrawn once the spouse reaches the normal retirement age.

In the death in retirement situation, the death benefit will allow the deceased retiree's surviving spouse to continue receiving pension benefits after the death of the retiree if the retiree chose a joint and survivor benefit payout option, and the pension benefit will continue until the death of the survivor. The size of the pension benefit received by the survivor depends on the payment option chosen by the retiree at the time of retirement, such as 100 percent or 50 percent. The 100 percent payment will allow the surviving spouse to receive 100 percent of the pension benefit paid to the retiree prior to death, whereas the 50 percent payment will entitle the surviving spouse to only 50 percent of the pension benefit received by the retiree. Since various payment options are made to be actuarially equivalent in terms of present value of total cost, the annual pension benefit received by the retiree under the 100 percent payment option should be smaller than that under the 50 percent payment option.

2.2.3 Health Benefits

Health benefits of retirees are turning out to be an increasingly important financing issue for state and local governments. Because many in the public sector retire before the age of 65, they are not eligible for Medicare. Most pension plans still pay for all or part of the retirees' health benefits until they become eligible for Medicare. Even then, some public plans pay for part of the Medicare premium or part of the prescription drug cost. Due to the substantial long-term cost of health benefits for retirees, it is a subject that will be discussed in detail in Chapter 8.

2.2.4 Withdrawal Benefits

When an employee terminates her employment, but does not have a vested right to a pension, she can withdraw all of her past contribution plus accumulated interest from the pension plan. As for the employer's contribution, some plans will not allow for withdrawal of any employer contribution prior to being vested. In other pension plans, there is a graduated vesting schedule. The employee will be entitled to a certain percentage of the employer's pension contribution plus interest depending on the number of years of service prior to full vesting. For example, if the full vesting period is five years, then for every year the employee works, she can be entitled to 20 percent of employer contribution. The main purpose of a withdrawal benefit is to make defined benefits more portable.

2.3 Deferred Retirement Option Plan (DROP) Program

In recent years, another type of pension benefit has become available to more and more state and local employees. It is called deferred retirement option plan (DROP). In this section, we look at the major features of DROP, the advantages and disadvantages of DROP for both employer and employee, and the major design issues concerning DROP.

2.3.1 Major Features of DROP

A DROP program contains many features that are tied to a normal retirement benefit.

1. **Participation requirement**: When a member of a pension plan meets the minimum age and/or service requirements for unreduced normal pension benefit, she can elect to join a DROP program, which will allow her to work in her current position and receive her regular salary while also receiving normal retirement benefits as if she is retired.
2. **Length of DROP program**: The term of the DROP program usually lasts three to five years. At the end of the DROP program, it is mandatory for the employee to retire from her position.
3. **DROP benefits:** The normal retirement benefit is calculated in the same way as if the employee has retired. It is deposited into an individual account and interest will be accrued on the account balance, with the interest rate either guaranteed by the plan sponsor, which is usually equal to the projected rate of investment return on pension assets, or based on the actual return of the retirement funds during her participation period. It is also possible there is no interest rate credit. However, the benefits the DROP participant receives are not available for use while the employee is in the DROP program.
4. **Employee and employer contribution**: Accrual of additional normal retirement benefit by a DROP participant is discontinued upon election into the DROP program, despite the fact that she continues to work in her position for a number of years. In other words, she is no longer considered an active member of the pension plan. Because of this, employee and employer contributions to the pension plan are also discontinued upon the employee's participation in the DROP program.
5. **DROP benefit payment options**: At the end of the DROP program, the employee will receive her DROP benefits. There are at least two major options to receiving this benefit. She can receive a lump sum of deferred retirement benefits plus interest accumulated during the DROP period. She may also elect to receive the DROP benefit in an annuity, which is a set amount either for a number of years or for life. The periodic payment amount depends on the length of the annuity. Distributions from DROP accounts may also be rolled over to another tax qualified plan, such as an individual retirement account.
6. **Retirement benefit after DROP**: At the end of the DROP program, in addition to receiving the DROP benefits, the employee also starts to receive her normal retirement benefits. The initial benefit amount she receives is the same as the one she receives on her participation in the DROP program. In other words, the retirement benefit has been frozen and not adjusted for inflation during her participation in DROP. Afterward, her retirement benefit will be increased periodically due to the cost of living adjustment. Since her retirement benefit has been frozen for several years, her initial annual benefit level upon final retirement is lower in that year than what she would have received had she retired permanently at the time she entered into DROP.

The Florida Retirement System (FRS) Pension Plan offers such a typical DROP program for up to five years. The member earns a salary and retirement benefit while in DROP. DROP accounts earn interest compounded monthly at an effective annual rate of 6.5 percent. She will earn no service credits while in DROP and her benefits will be calculated based on years of service at the time DROP participation begins. When the DROP period ends, she must terminate all employment with all FRS employers. She will receive a lump-sum payment of her accumulated DROP account and begin receiving monthly retirement benefits, determined at the time of DROP participation, plus an annual increase of 3 percent from then on.

2.3.2 Pros and Cons of DROP

The DROP program was first created in Baton Rouge, Louisiana, in the 1980s as an incentive for police officers and firefighters who were eligible for retirement to stay on the job for a few more years. In the following years, the DROPs created were mainly geared toward police and fire departments and the focus was mostly on retention. The reason that police officers and firefighters were targeted for retention is that, due to the demanding nature of their work, they are allowed to retire at a much younger age than other state and local government employees. In order to retain some of the police officers and firefighters for a few more years and also to better manage the timing of retirement, DROPs were created to keep the employees for an extra three to five years. Because they have to retire after the DROP period is over, the employer will know several years in advance the exact retirement date, allowing for better personnel management in terms of planning and projection. In addition, retention of older employees also defers any cost related to recruiting, hiring, and training of new employees.

Gradually DROP has become available to other state and local employees. This is especially important for government units that are experiencing and will experience a shortage of skilled workers. The retention of skilled senior employees can be a significant benefit of DROP. Based on a survey of state pension plans by the Office of New York State Comptroller (2006), 13 states had DROP programs in 2006, with 7 of these states limiting the DROP to law enforcement personnel and the other 6 providing the DROP to all employees.

Based on the brief history and the main features of DROP, some pros and cons of DROP to employers and employees are discussed below:

1. **Employer advantage**: A DROP can help employers retain valued employees who are eligible to retire with full benefits and who, otherwise, have little incentive to continue working. This will help public employers in retirement planning and ease the transition between the retiring employees and their successors. Because employer contributions to the defined benefit pension plan on behalf of an employee typically cease when the employee begins DROP participation, the public employer may realize a cost savings, when compared to replacing them immediately with new employees.
2. **Employer disadvantage**: The biggest potential disadvantage of DROP is that it may be more costly to the public employer than not having such a program. Despite the payroll savings mentioned earlier, true cost neutrality in DROP programs is difficult to measure and achieve, a subject discussed further in the last part of this section.
3. **Employee advantage**: DROP participation provides an opportunity for participants to accumulate and later receive a portion of their pension benefit in a lump sum upon final termination, while continuing to receive compensation for full-time employment. In essence, the account acts as a kind of employee savings plan. This is particularly beneficial if there is a cap on the benefit to be received, as is the case for several pension systems discussed earlier. This lump sum can further enhance income security during retirement.
4. **Employee disadvantage**: Because the retirement benefit is frozen at the time of DROP election, the employee who elects to participate in DROP agrees to forego an increase in benefit accruals while continuing to work. Such increase may come from three sources: (1) the periodic COLA, (2) the increase in one's salary, and (3) the improvements in the pension benefit formula itself that are periodically granted by the public employers. Therefore, the employee needs to determine whether the lump sum she receives will more than offset the loss due to a permanent reduction in retirement benefits.

2.3.3 DROP Design Issues

The most critical policy issue in designing DROP is the cost of such a program. While the payroll cost savings are part of the consideration, the more important part has to do with the retirement benefit itself. The objective in designing a DROP is to make it "cost neutral," meaning that total cost to the public employer should be the same whether the employee retires and starts collecting benefits or enters a DROP program on the day she plans to retire and defers her retirement for two to five years. There are two main factors in design that will have an impact on the eventual cost. First is the feature of the DROP. For each of the main features of a typical DROP discussed earlier, there can be many variations, such as whether interest should be credited to the DROP account and at what rate. Some variations are more costly than others. The cost implication of this first factor, however, can be estimated and controlled by the plan sponsor with relative ease. The second factor has to do with the assumption about retirement age. As will be discussed more fully in the next chapter, to project future normal retirement benefits, an assumption has to be made as to what age a plan member is going to retire to receive unreduced benefits. To make it actuarially neutral, the key assumption is that the decision to participate in a DROP does not change the original timing of retirement. If people retire earlier and enter into a DROP program, then they will collect benefits for a longer period of time and, thus, will cost the plan more. If an employee works longer because of DROP, it will actually lead to benefit savings for the employer. To know the cost to the public pension plan, the plan sponsor needs to compare the age of employees who enter into a DROP program to the age of those who retire without a DROP program. Therefore, the full cost of DROP may not be known for many years to come and it is also difficult to determine initially whether it is economical to implement a DROP.

2.4 Adequacy of Public Pension Benefits

The main objective of pension benefits is for the retirees to maintain the quality of life in retirement so that they can live in dignity. The adequacy of public pension benefits should be measured against this objective. While there is debate on what is considered the quality of life in retirement (as it can be highly subjective depending on the individual), a common, objective way of measuring the quality of life in retirement is that it is similar to the quality of life enjoyed by the individual prior to retirement. This means that the income in retirement should be comparable to the preretirement income in order to maintain the same quality of life. It is against this criterion that the adequacy of pension benefits in the public sector is evaluated.

This criterion is also captured in the concept of "replacement ratio," which is defined as the percentage of preretirement income that needs to be replaced by retirement income to produce an equivalent standard of living. The evaluation of the adequacy of pension benefits in the public sector involves two steps: (1) finding the replacement ratio to maintain the standard of living and (2) measuring public pension benefits against this ratio.

To analyze the adequacy of public pension benefits, we first start with an employee who retires at the age of 65 after 35 years of service in the public sector.* She also belongs to a pension plan that participates in the Social Security program and requires member contribution. Then we examine some complicating factors.

* We use the retirement age of 65 because it is the age at which full Social Security benefits can be collected. We also use 35 as the number of years for a full career because Social Security benefits are based on 35 years of earnings.

2.4.1 Finding the Replacement Ratio

Finding the replacement ratio involves analyzing what preretirement income is used for to see which part of that is no longer needed in retirement. What is left is the part that needs to be replaced. While there are several ways of arriving at a replacement ratio, some more complicated than others, the replacement ratio analysis in this chapter takes the simpler approach of only examining common factors that are observable, quantifiable, and apply to most retirees.

Preretirement income usage can be divided into three main categories: taxes, savings, and living expenses.

1. **Taxes**: Taxes that are dependent on the income earned are federal and state personal income taxes, and payroll tax for Social Security and Medicare. Since retirees no longer have to pay payroll tax at a combined tax rate of 7.65 percent, this alone means that 7.65 percent of preretirement income is no longer needed. One factor that will change the federal income tax amount in retirement is that the Social Security benefit, a part of total retirement income, is partially exempt from federal income tax. Generally, up to 50 percent of the Social Security benefit is taxable.* As will be seen in section 2.4.2, Social Security benefits replace 30 percent of preretirement income for public employees. Assuming a retiree is in the 15 percent marginal tax rate bracket, the reduction in federal income tax due to partial exemption of Social Security benefits amounts to 2.25 percent of one's preretirement income.† By combining the reductions in payroll tax and federal income tax, about 10 percent of preretirement income is no longer needed.‡
2. **Savings**: Savings here refer to employee pension contribution to the public pension plan. No personal savings is assumed in this analysis of replacement ratio. The retiree no longer has to contribute to the pension plan in retirement. For those pension systems that are under the Social Security coverage and also require employee pension contribution, the average employee pension contribution rate was five percent in 2004 based on the Wisconsin survey, meaning another five percent of preretirement income is no longer needed in retirement.
3. **Living expenses**: Compared to preretirement living expenses, work-related expenses will go down and age-related expenses will go up in retirement. The largest age-related expense is healthcare expenses and it naturally increases as

* U.S. Department of Treasury, Internal Revenue Service. Publication 915 "Social Security and Equivalent Railroad Retirement Benefits," for use in preparing 2006 return.

† Multiplying 30 percent by 50 percent and by 15 percent yields 2.25 percent.

‡ Since a public pension benefit is fully or partially exempt from state income tax in many states, the actual reduction in tax is even greater for many retirees in the public sector. However, as this tax treatment is not universally available, this factor is not considered in this analysis of replacement ratio.

people get older. Work-related expenses, such as food, clothing, and transportation, will decline in retirement. Whether the reduction in work-related expenses can offset the increase in age-related expenses will largely depend on individual circumstances. For the sake of simplicity, living expenses are assumed to be the same before and after retirement.

By combining the reductions in federal payroll and income taxes and employee pension contribution, a replacement ratio of 85 percent of preretirement income is adequate for a retiree in the public sector to maintain one's quality of life. In comparison, a more systematic study on replacement ratio by Aon Consulting (2004) found that the replacement ratio ranges from 78 to 89 percent for income levels that range from $20,000 to $90,000.

2.4.2 Adequacy of the Public Pension Benefit

Since the two main components of postretirement income for public retirees are Social Security benefits and public pension benefits, the share of Social Security benefits need to be subtracted from the replacement ratio before the adequacy of pension benefits can be analyzed. The formula that determines Social Security benefits makes replacing a higher percentage of income when the income level decreases. The monthly Social Security benefit is called primary insurance amount, or PIA. The PIA is determined by applying the Social Security benefit formula to a worker's average indexed monthly earnings (AIME). The AIME is the monthly average of a worker's 35 best years of earnings, with earnings before age 60 indexed to average wage growth. For workers who became eligible for benefits in 2004, PIA equals 90 percent of the first $612 of AIME, plus 32 percent of the next $3077 of AIME, plus 15 percent of AIME above $3689 (Government Accountability Office, 2004). Dividing the population into low, average, high, and maximum (the maximum earned income subject to Social Security tax) wage earners, the Social Security benefit's replacement ratios for these four groups are 49, 37, 30, and 24 percent, respectively. The cutoff points for these four groups are 45, 100, and 160 percent, respectively, of Social Security's Average Wage Index, which was $35,648 in 2004. Since the average salary for a state worker in 2004 was $41,118,* a public employee is between the average and high earner groups. However, because an average retiree's salary should be higher than an average state worker's salary, the retiree's salary should be closer to the high rather than the average cutoff point.

To simplify the calculation, we take the replacement ratios of 30 percent from Social Security benefits for public sector retirees. After subtracting 30 percent from the overall replacement ratio of 85 percent, that leaves 55 percent to be replaced

* U.S. Department of Labor, Bureau of Labor Statistics. Employment and Wages, Annual Averages 2004. Table 8. State government by state and selected industries: establishments, employment, and wages, 2004 annual averages. http://www.bls.gov/cew/ew04table8.pdf.

by public pension benefits. To determine whether public pension benefits meet this threshold, the pension benefit formula needs to be reexamined. Assuming an average working career of 35 years, it requires a multiplier of about 1.6 percent to reach the replacement ratio of 55 percent. Most of pension systems under the Social Security coverage adopt multipliers that are either 2 percent or close to 2 percent. Therefore, multiplying 2 percent by 35 years yields a replacement ratio of 70 percent, which is higher than what is considered adequate to maintain the quality of life. For pension systems with a multiplying factor greater than 2, such as 2.5 percent, then after 30 years of service, it is possible that the total pension benefit and Social Security benefit can exceed preretirement income.

While this analysis shows that most of the pension systems provide pension benefits at a level that is above the adequate level for retirees to maintain their quality of life when the Social Security benefits are taken into consideration, there are several factors that complicate this relatively simple analysis, which is based on the assumption that an employee retires at the age of 65 after 35 years of service and has Social Security coverage. In reality, there are deviations from this assumption in the public sector:

1. **Benefit formula**: Ideally, the pension benefit is based on the final salary. This is slightly complicated by the fact that final average salary used in determining pension benefit is an average of salaries in the final few years, rather than the final salary, and, thus, will be slightly smaller than the actual final salary. According to the Wisconsin Legislative Council (2005) survey, the average wage inflation in 2004 was 4 percent. Together with merit increase, salary increase should be at least 5 percent per year. Assuming a salary growth of 5 percent per year and the salaries of the final three years are used to determine the FAS, then the FAS is 5 percent less than the actual final salary. This means that using FAS as the base, the replacement ratio from public pension benefit should be 60 percent, rather than 55 percent.
2. **Median retirement age**: While the normal retirement age is assumed to be 65 in the analysis, the median retirement age in the United States since the late 1990s has been around 62 years of age (Gendell, 2001). If the public employee retires and draws Social Security benefits at the age of 62, there will be a reduction in the benefit as 65 is the age at which full Social Security benefits can be withdrawn. According to the Social Security Actuary's table on the effect of early retirement, a person who retires in 2007 at the age of 62, the earliest age at which Social Security benefits can be withdrawn, her benefit will be reduced by 25 percent.* For people who reach 62 in the next few years, the reduction will gradually increase to 30 percent. A 25 percent reduction will reduce the Social Security replacement ratio of 30 percent down to

* Social Security Administration. 2005. Effect of Early or Delayed Retirement on Retirement Benefits. http://www.ssa.gov/OACT/ProgData/ar_drc.html (Accessed May 31, 2007.)

about 22.5 percent. Had this public retiree withdrawn Social Security benefits at the age of 62, then about 63 percent of preretirement income needs to be replaced by public pension benefit. At the age of 62, a full career in the public sector may not last 35 years. Suppose the length of the full career is reduced by three years, then to reach a replacement ratio of 63 percent plus 5 percent to adjust for the FAS base, a multiplier of 2 percent is needed, which was the average of the multipliers used by major public pension plans in 2004. This means that for a public employee retiring at the age of 62 after a full career, the pension benefit plus Social Security benefit (even after a 25 percent reduction) is sufficient to maintain her quality of life.

3. **Social Security coverage**: As discussed in Chapter 1, about 25 percent of public employees are not covered by Social Security. Therefore, those who are not in the Social Security system should enjoy a higher level of public pension benefits. As the multiplier is 2.5 percent for most of the pension systems in this group, for an employee after a full career of 32 years as in the previous case, the replacement ratio is 80 percent, slightly below the adequate replacement ratio. Even though the employee in this group pays a higher pension contribution rate than a typical public worker in this analysis, the combined pension contribution rate and payroll tax rate is still much lower for the employee who does not have to pay the Social Security tax. Employees in this group then can use the additional savings to make up for the difference.
4. **Noncontributory systems**: As mentioned earlier, while most public pension plans require members to contribute, some plans are noncontributory and the benefit multiplier for these plans is also lower. The average benefit multiplier for these plans is close to 1.6 percent. At the age of 62, after 32 years of service, 51 percent of final average salary is replaced, below the 63 percent replacement ratio.

2.4.3 Summary

While the majority of employees in the public sector enjoy a public pension benefit that is adequate to maintain their quality of life if they retire at the age of 62 (after 32 years of service with Social Security coverage), this conclusion can be more difficult to draw for certain groups of public employees who either do not contribute to their pension plans or do not have Social Security coverage. It requires further analysis of what these public employees do with the part of their salary that they do not have to pay toward Social Security tax or pension contribution. If the amount is saved and invested every year for 30 years, they should have sufficient personal assets in retirement to raise their income replacement ratio to a level comparable to that of a typical worker in this analysis.*

References

Aon Consulting. 2004. *Replacement Ratio Study: A Measurement Tool for Retirement Planning*. http://www.aon.com/about/publications/pdf/issues/rs_2004_06_replacement_ratio_study_674.pdf (Accessed May 30, 2007.)

Gendell, Murray. 2001. Retirement age declines again in 1990s. *Monthly Labor Review*, October. http://www.bls.gov/opub/mlr/2001/10/art2full.pdf (Accessed May 31, 2007.)

Government Accountability Office. 2004. *Social Security: Distribution of Benefits and Taxes Relative to Earnings Levels*. (GAO-04-747), Washington, D.C.

Mcginty, Jo Craven, and David W. Chen. 2007. "Trying to Save, New Jersey Drove Up Pension Costs." *New York Times*, June 15, 2007, A1.

New Jersey Public Employees' Retirement System. 2005. Member Handbook. http://www.state.nj.us/treasury/pensions/epbam/exhibits/handbook/persbook.pdf (Accessed June 6, 2007.)

Office of the New York State Comptroller. 2006. *Report on Deferred Retirement Options and Partial Lump Sum Options*. Albany, NY.

Oklahoma Public Employees Retirement System. 2006. *State and Local Government Employees Member 2006 Handbook*. http://www.opers.state.ok.us/pdfs/regular_book_06.pdf (Accessed June 21, 2007.)

Pennsylvania Public Employee Retirement Commission. 1995. *Feasibility of Early Retirement Incentives in the Public Sector*. http://www.perc.state.pa.us/perc/lib/perc/1995_FEASIBILITY_OF_EARLY_RETIREMENT_INCENTIVES_IN_THE_PUBLIC_SECTOR.pdf (Accessed May 23, 2007.)

State Employees' Retirement System of Illinois. 2005. *Your Rights and Responsibilities: Employee Handbook*. http://www.state.il.us/srs/PDFILES/HANDBOOK.pdf (Accessed June 16, 2007.)

Wisconsin Legislative Council. 2005. *2004 Comparative Study of Major Public Employee Retirement Systems*. http://www.legis.state.wi.us/lc/publications/crs/2004_retirement.pdf (Accessed June 6, 2006.)

Wisconsin Retirement System. 2005. *Comprehensive Annual Financial Report for the Fiscal Year Ended June 30, 2003*. http://elf.wi.gov/about/2003_cafr.pdf (Accessed June 4, 2006.)

* A simple example will make this point clear. Suppose an employee retires at the age of 62 with a final salary of $50,000. To bring the replacement ratio of 51 percent to 63 percent, then an additional $6000 is needed every year for 21 years, assuming a life expectancy of 83 at the age of 62. At a discount rate of 6 percent, the total amount needed at the age of 62 is $70,000. If he saves $800 a year for 32 years with an investment return of 6 percent, he will have about $73,000 by the time he retires; $800 a year is less than the 5 percent employee contribution rate, even in early years. If the retirement benefit has to increase by 3 percent a year, then the annual savings needs to increase to $1000, still less than 5 percent of salary.

Chapter 3

Actuarial Valuation and Financial Reporting

There are two basic ways of funding defined pension benefits: pay-as-you-go and advance funding (also called prefunding), as discussed in Chapter 1. Compared with advance funding, pay-as-you-go funding has two key disadvantages: escalating pensions cost and lack of investment income. First, since the ratio of retirees to active members of a pension plan tends to increase with the aging of population, pay-as-you-go leads to escalating pension cost in the future and puts tremendous pressure on the plan sponsor's budget. Second, since no funds are set aside and invested, no investment income is earned. As investment income is a far more important source of funding over time to pay for pension benefits, a lack of investment income will lead to a much greater employer and employee pension contribution over time compared to advance funding. Even if pay-as-you-go can be less costly than advance funding in the short run, it can be far more expensive in the long run. Because of these major disadvantages, almost all public pension plans are now prefunded.

To prefund pension benefits, an actuarial valuation of the pension plan is a prerequisite. Actuarial valuation is the process that determines the long-term cost of a pension plan, the necessary contribution rate and the required return on investments to accumulate sufficient assets over time. It also evaluates the sufficiency of the assets accumulated in comparison to pension benefits already earned by employees at any given time. The actuarial valuation, arguably the most technical part of public pension plan management, is a fundamental test of the long-term financial viability of a pension plan. This chapter introduces the basic concepts and procedures involved in actuarial valuation. This will help us understand not only the financial reports on public pension plans, as discussed later in this chapter, but also the significance of various public policies guiding the management of public pension plans, as will be discussed in Chapter 6.

3.1 Assumptions

An actuarial valuation is based on a set of assumptions on demographic and economic parameters. Due to the long-term nature of pension benefits, these assumptions play a critical role in determining the total cost, the required contribution rate, and the sufficiency of pension plan funding.

3.1.1 Demographic Assumptions

The demographic assumptions are concerned with parameters that cause a decrease in the plan membership and retiree population, due to mortality, disability, and termination.

The mortality rate refers to the probability that people will live to various ages. It concerns both active members and retirees. The mortality rate of active members refers to the probability the member will survive to retirement age to collect pension benefits. The mortality rate of retirees projects how long they will withdraw pension benefits.

The disability rate refers to the probability employees could become disabled and collect disability benefits. Since disability benefits are usually more expensive than normal service benefits, especially when an employee is disabled relatively early in his career, a higher disability rate will lead to overall higher pension benefit liability for the pension plan.

The termination rate refers to the turnover rate for active members of the plan. If a member terminates employment prior to vesting, he forfeits his right to future pension benefits, although he is entitled to a refund of his contributions. If a member terminates employment after vesting, he will be entitled to a deferred pension benefit, which will be frozen at the time of termination. Either way, termination reduces a plan sponsor's future pension benefit liability.

The sum of various decremental rates among active members is also called survival rate. It refers to the probability an active member will survive in employment until retirement age. The lower the survival rate, the fewer the people who will collect pension benefit and the smaller the future pension liability. This survival rate increases with the increase in a plan member's age. For example, when an employee gets older, especially after vesting, the probability of him terminating employment will gradually decrease.

Implicit in the mortality rate assumption is also the assumption about the retirement age, which is the probability of employees retiring at various ages. Such assumption is largely based on a plan's past experience. Along with the mortality rate, the assumption on retirement age will determine how long a retiree will withdraw pension benefits. An increase (decrease) in average retirement age will decrease (increase) future pension liabilities.

3.1.2 Economic Assumptions

The key economic assumptions that have to be made for actuarial valuation are inflation rate, cost of living adjustment, salary increase, and rate of investment return. Of these four, the rate of return is by far the most important assumption made in actuarial valuation.

The rate of return assumption reflects what average return can be earned on pension asset investment over a long period of time. The higher the rate, the greater the amount of investment income will be earned and the less the pension contributions are needed from both the employer and employee to fund future pension benefits. More importantly, the rate of return, sometimes referred to as interest rate, is also used as a discount rate. It is the rate used to discount all future pension benefits to the present value on the date of valuation. The present value of future pension liabilities is used in the determination of the sufficiency of a pension plan's funding level. Everything else being equal, the greater the discount rate, the smaller the present value of future liabilities, and the higher a pension plan's funding level and, thus, the less the pension contribution needed. It has been found that a change of 1 percent in the rate of return assumption alters the long-run cost estimate by about 25 percent (Bizley, 1950). Therefore, an assumption of a lower rate of return generally indicates a more conservative approach to actuarial valuation, everything else being equal. Of the 83 large public pension plans in the Wisconsin Legislative Council's 2004 Comparative Study of Major Public Employee Retirement Systems survey, the mean and median interest rate was 8 percent. It was used by 39 plans, and another 33 plans used an interest rate that was within a half percentage above or below the median rate.

The rate of return assumption also includes an assumption about inflation rate because the rate of return has two components: the real rate of return and the inflation rate. The average inflation rate assumption in the same survey in 2004 was about 4 percent, with a range from 3 to 6 percent. While a higher inflation rate assumption can lead to a higher rate of return assumption, it does not necessarily mean a less conservative approach to actuarial valuation, primarily due to its effect on other economic assumptions. First, it has a direct impact on the cost of living assumption. Many public pension plans have automatic cost of living adjustment (COLA) for retirees' pension benefits, as discussed in Chapter 2. The COLA is tied to the inflation rate assumption. A higher inflation rate assumption means a larger COLA adjustment and, thus, larger pension liability.

The inflation rate assumption also has a direct effect on salary progression. Salary progression assumption is important because an employee's future benefit level is based on this assumption, as can be seen from the defined benefit formula discussed in Chapter 2. Salary progression consists of two parts: merit increase and inflation increase. Merit increase is due to promotion and increased responsibility over one's working career. Like the COLA for retirees, employees' salaries also increase due to inflation. A higher inflation rate assumption thus leads to a higher benefit level, which requires higher pension contribution. In all, the inflation rate assumption has an offsetting effect on both asset and liability growth. A higher inflation rate means higher asset growth, but it is offset by higher pension liability growth, and vice versa.

3.2 Actuarial Cost Methods

Actuarial cost method is a procedure that determines future pension benefits, the portion of future pension benefits that should be allocated to each year for funding purpose, the pension benefits accrued to date, and how unfunded accrued pension liability, if it occurs, should be amortized. There are special actuarial terms that are associated with these four parts of actuarial valuation, briefly explained as follows:

1. Through projection, future pension benefits are first determined, which is then used to calculate the "actuarial present value of future benefits" at the date of actuarial valuation. It is so called because the derivation involves population decremental factors, such as termination and disability, and discount for the time value of money.
2. The portion of actuarial present value of future benefits assigned to a particular year is called "the normal cost." This is the cost to fund the pension benefit earned by a plan member due to the service provided in that year.
3. The present value of future benefits is divided into two parts. The first part is for services already rendered in the past and, thus, benefits have already been accrued. This is called "accrued actuarial liability," which is theoretically equal to the cumulative normal cost plus interest. The second part is equal to the present value of future benefits yet to be earned, which is equal to the present value of all future normal cost.
4. The portion of the accrued actuarial liability that is offset by pension assets already accumulated is called "funded actuarial liability." If the accrued actuarial liability is less then the value of pension assets, the difference is called "unfunded accrued actuarial liability" (UAAL). When a UAAL occurs, it needs to be amortized (or paid off) over a number of years. In this case, the annual pension contribution includes both normal cost and amortization cost.

3.2.1 Classification of Actuarial Cost Methods*

There are many ways of classifying actuarial cost methods, depending on the characteristics that define a cost method. The most important characteristic is whether the method allocates the benefits or the cost of benefits to various plan years. When the annual benefit is first allocated to the various plan years and then the cost related to the benefit is determined, such method is called the "benefit allocation method." If the present value of all future benefits (in other words, the cost of future benefits) is allocated to the plan years without first allocating the benefits, this is called "cost allocation method."

A second characteristic of actuarial cost method is whether the normal cost and actuarial liability for the plan are determined on the basis of each individual or the entire active plan participants. When done on an individual basis, the cost and liability for each individual are determined first and then are summed up for the entire group. When done on a group basis, there is no separate calculation for each individual's cost and liability.

A third characteristic of actuarial cost method is whether the cost and accrued liability of the plan are calculated with regard to the entry ages or the attained ages of plan participants. Entry age refers to a participant's actual or assumed age at which he enters into the service, whereas attained age refers to the participant's age at the time of actuarial valuation. For the entry age method, the present value of future benefits is allocated over a period between the entry age and the final assumed retirement age of a plan participant. For the attained age method, the present value of future benefits is allocated over a period between the attained age and the final assumed retirement age.

Different combinations of these major attributes lead to many different actuarial cost methods. Governmental Accounting Standards Board (GASB), which sets standards for financial accounting and reporting for public sector entities, finds acceptable six major actuarial cost methods for public pension plans:

- Unit credit (projected or unprojected)
- Entry age normal
- Attained age
- Aggregate cost
- Frozen entry age
- Frozen attained age actuarial cost methods

---

* The author would like to thank Dr. Buqing Yang from the Department of Insurance, College of Finance, Shanghai University of Finance and Economics, who has provided substantial technical assistance in providing the technical expertise on actuarial valuation and in designing the actuarial example in this chapter.

Table 3.1 Actuarial Cost Methods and Their Usage

Actuarial Cost Method	*Main Characteristics of the Cost Method*	*Number of Plans*
Projected unit credits	Benefit allocation on an individual basis with separate unfunded liability	12
Entry age normal	Cost allocation on an individual basis with separate unfunded liability	60
Aggregate	Cost allocation on a group basis with no separate unfunded liability	5
Frozen entry age	Cost allocation on a group basis with frozen initial unfunded liability	5

Source: Wisconsin Legislative Council 2004 Comparative Study of Major Public Employee Retirement Systems.

Based on the 2004 Wisconsin survey, four of these six cost methods are used by the 83 large plans in the survey. Table 3.1 shows the usage of these four methods.*

The vast majority of pension plans use an actuarial cost method that allocates cost rather than benefit over the service years and calculates pension liability on an individual basis rather than a group basis. Of these four methods, entry age normal is by far the most used cost method with about three-quarters of the pension plans adopting it. Following is a detailed discussion of these four actuarial cost methods in terms of the recognition of normal cost, accrued actuarial liability, and unfunded accrued actuarial liability. Given the importance of entry age normal (EAN) and projected unit credit (PUC) cost methods, a comparison will also be made between the two.

3.2.2 Projected Unit Credit (PUC) Cost Method

Projected unit credit (PUC) is a cost method that allocates benefit rather than cost of benefit to the service years of plan participants. It is also conducted on an individual basis rather than on an entire group.

3.2.2.1 Normal Cost

Under this method, the calculation of annual normal cost involves three steps. First, the participant's final salary at retirement can be determined through salary projection. The annual benefit then can be determined by using the benefit formula

* Since the survey, Oregon Public Employees Retirement System switched from entry age normal to projected unit credit cost method in December 2004, and Vermont State Teachers Retirement System switched from frozen entry age to entry age normal cost method.

involving the final salary, number of years of service, and a multiplying factor. Second, the annual pension benefit is allocated to each service year between the entry (or attained) age and final retirement age. This is accomplished by dividing the prospective annual retirement benefit by the number of years for which benefits are credited. In the third and final step, the actuarial present value of the pension benefit allocated to each year, which is also the normal cost for each year, is calculated based on the following formula:

Normal cost = Benefit allocated to each year × Survival rate from attained age to retirement age × Discounting factor from attained age to retirement age × Annuity factor of $1 paid annually at retirement age*

The survival rate, as discussed earlier, takes into consideration various decremental factors that will reduce the membership. The discount rate is the plan's assumed rate of return. With each passing year, the normal cost increases due to these two factors: (1) the survival in service rate increases as a participant gets closer to final retirement and (2) the discount factor also increases due to the shorter period for discounting. The annuity factor, however, does not change from year to year. The combination of the increase in survival rate and the discount factor means a substantial increase in normal cost from the earlier years to the later years.

3.2.2.2 Accrued Liability

As for the accrued actuarial liability, it is calculated as follows:

Accrued actuarial liability = Benefits allocated to date × Survival rate from attained age to retirement age × Discounting factor from attained age to retirement age × Annuity of $1 paid annually at retirement age

As in the case of normal cost, because both the survival rate and the discounting factor increase each year, liability is accrued at a fairly slow pace in the early years, but increases at a much faster pace in the later years.

While theoretically it is possible to allocate benefits to each service year on a level percentage of salary rather than on a level dollar basis, the initial normal cost will be even smaller and, thus, pension plans are not allowed to use level percentage allocation for the PUC cost method.

---

* The annuity factor is the present value of $1 received each year from retirement age until death. It is dependent on two main factors, the number of years the retiree will receive the annuity and the interest rate for discounting the future annuity payments. The number of years receiving the annuity payment is determined by the group mortality and the discount rate should be the same as the assumed interest rate of the pension plan.

3.2.2.3 *Example*

Following is a simple example to show how annual normal cost and accrued liability are derived at under the PUC cost method. The example is based on the following assumptions:

Entry age: 30
Retirement age: 65
Salary at entry age: $30,000
Salary increase: 5 percent per year
Discount rate: 8 percent
Pension benefit: 50 percent of final year's salary

Table 3.2 shows the derivation of annual normal cost and accrued liability between entry age of 30 and retirement age of 65.* First, to calculate the normal cost, annual pension benefit needs to be determined. Through salary projection, the final salary at retirement is $165,480, and the annual benefit at retirement is $82,740. Divided by 35 years of service, the benefit allocated to each year of service is $2364. Multiplying the annual benefit by the survival rate and the discount factor and by the annuity factor of $1 yields the annual normal cost.† Because both the survival rate and discount factor increase every year, the normal cost also increases every year. When measured against the annual projected salary, annual normal cost under PUC accounts for a much higher percentage of salary in later years than in early years. This is also shown in accrued liability.‡ It increases very slowly in the early years, but at a much faster pace in later years.

3.2.2.4 *Unfunded Accrual Actuarial Liability (UAAL) and Amortization*

By comparing the accrued actuarial liability to assets already accumulated in the pension plan, the funding status can be determined using a funding ratio. The funding ratio is calculated by dividing the value of pension assets into accrued actuarial liability. When the asset value is equal to or greater than the accrued liability, the ratio is equal to or greater than 100 percent. When the asset value is less than the accrued liability, the funding ratio is less than 100 percent and a UAAL occurs

---

* The mortality rate is from China's 2000 National Urban Male Mortality Rate Table. The withdrawal rate is from China's Individual Annuity Lapse Table. The mortality rate and withdrawal rate together determine the survival rate from attained age to retirement age of 65. Based on the mortality rate, the annuity factor of $1 at age 65 is 8.64.

† For example, at the age of 30, the normal cost of $166.72 = $2364 × 0.1206 × 0.0676 × 8.64.

‡ As an example of calculating accrued liability, at the age of 31, after two years of service, the accrued liability of $445.16 = 2 × $2364 × 0.1491 × 0.0730 × 8.64.

Table 3.2 Normal Cost and Accrued Liability under Projected Unit Credit (PUC) Cost Method

Age	*Mortality Rate*	*With-drawal Rate*	*Survival Rate to Age 65*	*Discount Factor*	*Normal Cost Contribution*	*Normal Cost as Percentage of Salary*	*Accrued Liability*
30	0.00104	0.190	0.1206	0.0676	166.72	0.00556	166.72
31	0.00096	0.120	0.1491	0.0730	222.58	0.00707	445.16
32	0.00105	0.095	0.1697	0.0789	273.16	0.00827	820.29
33	0.00103	0.085	0.1877	0.0852	326.72	0.00941	1306.89
34	0.00117	0.075	0.2054	0.0920	386.07	0.01059	1930.36
35	0.00129	0.070	0.2223	0.0994	451.34	0.01179	2708.02
36	0.00134	0.060	0.2394	0.1073	524.86	0.01306	3674.03
37	0.00125	0.055	0.2550	0.1159	603.89	0.01431	4831.14
38	0.00162	0.050	0.2703	0.1252	691.28	0.01560	6221.48
39	0.00183	0.050	0.2850	0.1352	787.21	0.01691	7872.13
40	0.00211	0.045	0.3006	0.1460	896.66	0.01835	9863.31
41	0.00198	0.045	0.3154	0.1577	1016.27	0.01981	12,195.29
42	0.00235	0.040	0.3310	0.1703	1151.68	0.02138	14,971.87
43	0.00235	0.040	0.3456	0.1839	1298.82	0.02296	18,183.51
44	0.00253	0.040	0.3609	0.1987	1464.76	0.02466	21,971.41
45	0.00285	0.040	0.3769	0.2145	1652.21	0.0269	26,435.36
46	0.00304	0.040	0.3938	0.2317	1864.27	0.02847	31,692.60
47	0.00333	0.040	0.4115	0.2502	2103.97	0.03060	37,871.41
48	0.00365	0.040	0.4301	02703	2375.20	0.03290	45,128.83
49	0.00391	0.040	0.4498	0.2919	2682.30	0.02528	53,646.01
50	0.00458	0.040	0.4704	0.3152	3029.93	0.03806	63,628.50
51	0.00459	0.040	0.4924	0.3405	3425.01	0.04098	73,350.22
52	0.00513	0.040	0.5154	0.3677	3871.65	0.04412	89,047.89
53	0.00549	0.040	0.5397	0.3971	4379.00	0.04752	105,096.09
54	0.00625	0.040	0.5654	0.4289	4954.71	0.05121	123,867.84
55	0.00682	0.040	0.5929	0.4632	5610.58	0.05523	145,875.09
56	0.00701	0.040	0.6220	0.5002	6357.06	0.05960	171,640.74
57	0.00827	0.040	0.6527	0.5403	7204.30	0.06432	201,720.51
58	0.00874	0.040	0.6858	0.5835	5175.27	0.06952	237,082.78
59	0.01001	0.040	0.7209	0.6302	9281.68	0.07516	278,450.37
60	0.01155	0.040	0.7588	0.6806	10,551.91	0.08138	327,109.35

Table 3.2 Normal Cost and Accrued Liability under Projected Unit Credit (PUC) Cost Method (Continued)

Age	*Mortality Rate*	*With-drawal Rate*	*Survival Rate to Age 65*	*Discount Factor*	*Normal Cost Contribution*	*Normal Cost as Percentage of Salary*	*Accrued Liability*
61	0.01209	0.040	0.8001	0.7350	12,015.46	0.08826	384,494.87
62	0.01359	0.040	0.8441	0.7938	13,689.80	0.09577	451,763.53
63	0.01474	0.040	0.8919	0.8573	15,622.18	0.10408	531,154.15
64	0.01625	0.040	0.9435	0.9259	17,849.01	0.11325	642,715.35
65			1.0000	1.0000	20,431.30	0.12350	715,095.50

that needs to be amortized over a period of time. The amortization cost each year includes both the principal amount of UAAL and the interest on UAAL. There is interest on UAAL because it represents the investment income that could have been earned had the asset been in the pension plan to be invested. The interest rate on UAAL is equal to the assumed rate of return of the pension plan.

The amortization cost can be calculated either as level dollar or as level percentage of payroll. The amortization cost under the level percentage method is smaller than that under the level dollar method in the beginning, but greater later on. The purpose of using the level percentage method is to reduce the initial burden of paying off the unfunded liability and to create a steadier contribution rate for the plan sponsor. As payroll increases, the amortization cost also increases. As the initial amortization cost under the level percentage method is usually smaller than the interest owed on the UAAL, the UAAL will continue to grow until the amortization cost is greater than the interest owed on the UAAL.

3.2.3 Entry Age Normal (EAN) Actuarial Cost Method

Entry age normal (EAN) is a cost allocation method performed on an individual participant using the participant's entry age as the starting point for cost allocation. Through the use of salary projections, the participant's total prospective pension benefit is estimated and then the actuarial present value of that benefit at the participant's entry age is determined. This value (or cost) is then allocated to each year of the participant's service years from entry age to retirement age in an amount that is either level dollar or level percentage of the participant's estimated salary from entry age to final retirement age. The derivation of a level dollar normal cost is discussed first and then the level percentage method is compared to the level dollar method.

3.2.3.1 Normal Cost under Level Dollar Method

The first step in determining the normal cost under a level dollar method is to estimate the participant's annual retirement benefit. The total retirement benefit (which is also the total cost to the employer) at the time of retirement then can be determined by multiplying the annual benefit by an annuity factor of $1 at retirement age.

In the second step, the actuarial present value of total cost at entry age is determined, which involves a population decremental factor and a present value discounting factor, the same two factors as discussed under the PUC cost method.

In the third and final step, the present value of total pension cost is to be allocated to service years on a level dollar basis. This involves calculating the temporary annuity factor of $1 contributed each year from the entry age until retirement age.* The actuarial present value of total pension cost is divided by the value of the temporary annuity of $1 and the result is the normal cost for this participant. Unless there is change to the pension benefit design or to the underlying actuarial assumptions, this normal cost will remain fixed until the participant's retirement age.

3.2.3.2 Accrued Liability under Level Dollar Method

Once the normal cost is determined, the participant's actuarial accrued liability can also be determined each year. It is calculated by subtracting the present value (PV) of future normal cost from the present value of future benefits, as seen in the following equation:

$$\text{Accrued liability} = \text{PV of future benefit} - \text{PV of future normal cost}$$

The accrued liability increases each year as the first term on the right-hand side of the equation becomes greater and the second term becomes smaller each year. Even though the total future pension benefit remains the same each year, the present value of total benefits increases every year because, with each passing year, the probability of the participant surviving in employment until final retirement age becomes greater and the time period over which to discount the total pension cost becomes shorter, resulting in larger present value. While the normal cost remains the same each year, the present value of future normal cost becomes smaller because there are fewer normal cost payments to make with each passing year.

The calculation of UAAL and its amortization are done in the same way as under the PUC cost method.

* Again, this annuity factor is the present value of $1 contributed each year from the date of valuation until final retirement age.

3.2.3.3 Entry Age Normal (EAN) Using Level Percentage

To calculate the normal cost under the level percentage method, the first two steps, calculating the total pension cost at time of retirement and then the present value of total cost at entry age, are the same as those for the level dollar method. The difference lies in the allocation of the present value over service years. Under the level percentage method, the actuarial present value of total salary earnings over service years is calculated. Dividing the present value of future pension liability by the present value of future earnings yields a certain percentage. Multiplying this level percentage by each year's annual salary produces the normal cost for that year. The actuarial accrued liability under the level percentage method is generated in the same way as under the level dollar method.

Since salary gradually increases, the normal cost under the level percentage method has to gradually increase as well. This means that the normal cost generated under level percentage method is smaller (greater) than that generated under the level dollar method in the earlier (later) years. Because of the smaller normal cost in the earlier years, the liability accrued under the level dollar method is also smaller than that under the level dollar amount in the earlier years. The advantage of level percentage over level dollar method is it creates a more stable contribution rate to the plan sponsor and makes it easier to budget, especially in the early years when a level dollar normal cost can loom especially large.

3.2.3.4 Example

Table 3.3 shows the normal cost and accrued liability using the EAN cost method. The basic assumptions in the EAN cost method example are the same as those in the previous PUC example. To calculate the normal cost under the level dollar method, the actuarial present value of future pension benefit is determined first, which at the entry age of 30 is $5835.* Dividing this number by the temporary annuity factor of $1, which is 6.25 at the age of 30, yields a normal cost of $933.10. This is the same normal cost for the remaining 34 years. Since the normal cost is level, it accounts for a decreasing percentage of annual salary, which increases every year. The accrued liability is derived by subtracting the present value of future normal cost from the actuarial present value of future benefits.

Under the level percentage method, first the level percentage needs to be calculated. The actuarial present value of all future salary is determined to be $273,812. Dividing $5835 by $273,812 yields 2.13 percent. Multiplying this percentage by the annual salary yields the annual normal cost. The accrued liability is achieved in the same way as under the level dollar method.

* $5835 = $82,740 × 0.12065 × 0.0676 × 8.64.

Table 3.3 Normal Cost and Accrued Liability under Entry Age Normal (EAN) Cost Method

	Level Dollar Method			*Level Percentage Method*	
Age	*Normal Cost Contribution*	*Normal Cost as Percentage of Salary*	*Accrued Liability*	*Normal Cost Contribution*	*Accrued Liability*
30	933.10	0.03110	933.10	639.33	639.33
31	933.10	0.02962	2178.84	671.29	1356.36
32	933.10	0.02821	3610.06	704.86	2303.15
33	933.10	0.02687	5246.24	740.10	3458.08
34	933.10	0.02559	7132.36	777.11	4850.13
35	933.10	0.02437	9271.17	815.96	6487.59
36	933.10	0.02321	11,714.58	856.76	8414.97
37	933.10	0.02210	14,411.62	899.60	10,603.06
38	933.10	0.02105	17,430.06	944.58	13,110.59
39	933.10	0.02005	20,782.17	991.81	15,956.42
40	933.10	0.01909	24,604.75	1041.40	19,259.03
41	933.10	0.01819	28,819.99	1093.47	22,968.31
42	933.10	0.01732	33,593.05	1148.14	27,230.68
43	933.10	0.01649	38,818.03	1205.55	31,971.82
44	933.10	0.01571	44,710.55	1265.82	37,385.69
45	933.10	0.01496	51,365.38	1329.12	43,569.45
46	933.10	0.01425	58,891.22	1395.57	50,634.76
47	933.10	0.01357	67,396.19	1465.35	58,695.33
48	933.10	0.01292	77,017.74	1538.62	67,893.27
49	933.10	0.01231	877,908.75	1615.55	78,387.44
50	933.10	0.01172	100,234.89	1696.33	90,351.35
51	933.10	0.01116	114,237.91	1781.14	104,031.23
52	933.10	0.01063	130,068.18	1870.20	119,593.34
53	933.10	0.01013	148,045.94	1963.71	137,364.28
54	933.10	0.00964	168,422.74	2061.89	157,630.32
55	933.10	0.00918	191,672.98	2164.99	180,816.79
56	933.10	0.00875	218,108.05	2273.24	207,314.12
57	933.10	0.00833	248,109.56	2386.90	237,507.84
58	933.10	0.00793	282,481.79	2506.25	272,212.13
59	933.10	0.00756	321,644.93	2631.56	311,890.39

Table 3.3 Normal Cost and Accrued Liability under Entry Age Normal (EAN) Cost Method

	Level Dollar Method			*Level Percentage Method*	
Age	*Normal Cost Contribution*	*Normal Cost as Percentage of Salary*	*Accrued Liability*	*Normal Cost Contribution*	*Accrued Liability*
60	933.10	0.00720	366,596.45	2763.14	357,555.49
61	933.10	0.00685	418,376.47	2901.29	410,285.34
62	933.10	0.00653	477,609.78	3046.36	470,754.88
63	933.10	0.00622	545,959.64	3198.68	540,672.20
64	933.10	0.00592	624,715.35	3358.61	621,388.12
65	933.10	0.00560	715,095.48	3526.54	715,095.48

3.2.4 Comparison between EAN and PUC

Because EAN and PUC are the two actuarial cost methods used most by public pension plans, it is useful to compare them in terms of growth in normal cost and accrued liability.

In terms of similarity, all three methods yield exactly the same final pension liability of $715,095 and this liability will be fully funded at final retirement under all three methods. The difference lies in the pace at which the liability accrues. Normal cost under PUC is much smaller than under EAN in early years. This can be seen in Figure 3.1. As a result, liability also accrues at a much slower pace under PUC in early years than under EAN, which is illustrated in Figure 3.2. Because of the much smaller normal cost and accrued liability in early years under PUC,

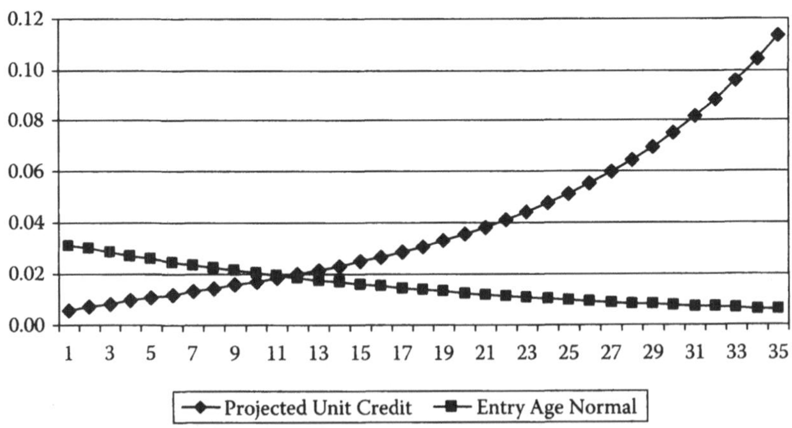

Figure 3.1 Normal cost as a percentage of salary under projected unit credit (PUC) and entry age normal (EAN) cost methods.

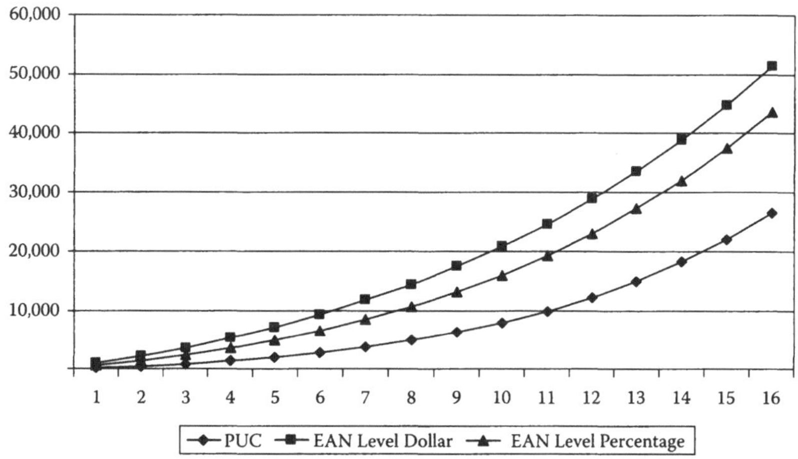

Figure 3.2 Accrued liability growth under various actuarial cost methods.

they have to increase at a much faster pace in later years when compared to those under EAN.

3.2.5 *Aggregate Cost Method*

The aggregate cost method is conceptually very similar to the entry age normal cost method with one major difference. Like EAN, the aggregate cost method allocates cost rather than benefits over the service years of plan members, and the cost allocated to the valuation year is the normal cost. The difference between the two cost methods is that the normal cost is calculated on an individual member basis under the EAN, whereas it is calculated for the entire group under the aggregate cost. The major consequence of this difference is in the recognition of UAAL. Due to the importance of UAAL in understanding the difference between these two methods, a more systematic understanding is needed of how UAAL occurs.

3.2.5.1 UAAL

There are four primary factors that can lead to UAAL. The first most important factor is the credit granted to services provided before the pension plan was established. In this case, calculating accrued liability based on entry age rather than attained age automatically generates a UAAL because no normal cost has been set aside for past service and, thus, no assets have been accumulated. As the UAAL at the inception of a pension plan can be fairly large, the sponsor of the plan has to amortize it over a period of time. This usually is not an issue when the membership of the plan at its inception is growing and the number of retirees is still small.

The second important factor is pension benefit increase. When a plan sponsor increases pension benefits, such increase is usually granted retroactively to past services. For example, if the multiplying factor is increased from 1.5 to 1.6 percent, that 1.6 percent will be applied to all past and future years of service rather than just to the years of service after the increase. This will create a UAAL as past contribution and asset accumulation is based on the lower pension benefit level. If the pension plan is overfunded at the time of pension benefit increase, it is possible that the surplus may offset the increase in pension liability and, thus, no UAAL is created.

The third important factor is unfavorable actuarial experience. Since the actuarial valuation is based on many assumptions, if the actual experience of the plan is unfavorable when compared to the assumptions, UAAL also occurs. Unfavorable actuarial experience can come from both the liability and asset sides. From the liability side, if the actual salary increase is greater than the underlying assumption, this leads to a pension liability greater than initially projected. Other factors include higher life expectancy, lower turnover rate, and earlier retirement age than assumed. From the asset side, the assumed rate of return is the most important factor in projecting how fast the pension assets will grow in the future to meet pension liability growth. If the actual investment return is below the assumed rate of return, the asset growth will fall short of the liability growth and, thus, a UAAL will occur. However, it is also entirely possible that the actual plan experience is more favorable when compared to the assumptions, such as higher investment return than the assumed rate. In such a situation, an "actuarial gain" is created.

The fourth important factor is pension contribution deficiency. Even if a pension plan has no initial unfunded liability, no pension benefit increase, and all the actuarial assumptions are met, the plan will stay fully funded only if the plan sponsor has made contributions equal to the normal cost every year. Any contribution less than the actuarially required amount can lead to slower asset growth and, thus, UAAL.

3.2.5.2 Aggregate Cost Method

Under EAN, after the normal cost is calculated, the difference between the PV of future benefit and the PV of future normal cost is the accrued liability. The part of the accrued liability that is not offset by pension assets is the unfunded liability, which is to be amortized. This unfunded liability can be further increased or decreased by future actuarial gain or loss. Under EAN, annual pension contribution consists of normal cost and amortization cost. Under the aggregate cost method, the normal cost for the entire group is calculated in such a way that the PV of future benefits is always equal to the present value of future normal cost plus pension assets. In other words, no UAAL is identified separately under the aggregate cost. Any unfunded pension liability due to past service benefits, retroactive pension benefit increases, or any actuarial gain or loss is automatically amortized over the remaining service years of active plan participants. The annual pension cost under the aggregate method thus makes no distinction between normal cost and amortization cost, as is done under the individual-based cost method. Viewed from another perspective, under EAN, the normal cost is fixed, but the amortization cost is variable. Under the aggregate cost, the entire normal cost is variable as there is no amortization cost. Since UAAL is not separately identified, the funding ratio for pension plans using an aggregate cost method is always equal or very close to 100 percent, even at the inception of a pension plan.

Despite this difference, these two methods are still conceptually very similar as both of them allocate cost over the service years of active plan participants. If the UAAL is amortized over a 20- to 30-year period of time under EAN, the total normal and amortization cost under EAN should also be fairly close to the normal cost under the aggregate cost method. The difficulty in the meantime, however, lies in assessing whether the aggregate cost plans have accumulated sufficient assets at any given time or if they are making progress toward a full funding level.

3.2.6 Frozen Entry Age Cost Method

This method is substantially similar to the previous aggregate cost method, with one major difference. Like the aggregate cost, it also allocates cost rather than benefit over the remaining service years of the plan members as a group. The major difference between this and the aggregate cost method is in the recognition of the initial unfunded liability. The aggregate cost does not recognize any UAAL, whether it is created at the inception of the pension plan or afterward. Any UAAL is automatically amortized over the remaining years of the members. Frozen entry age method, however, makes a distinction between UAAL created at the inception of the plan and those created later. The initial UAAL is calculated by using the entry age normal cost method and amortized over a period of time.* The initial liability is frozen at that level. After the plan inception, UAAL created by subsequent actuarial gain or loss or pension benefit increase will not be recognized separately under this method and will also be automatically amortized over the remaining service years of plan members, thus resulting in an increase in normal cost.

Because of the existence of the initial unfunded liability, the funding ratio will not be 100 percent. However, because the initial liability is frozen and is being amortized and because the plan assets grow over time, the funding ratio should show a steady increase as well. When the initial liability is largely paid off, then the plan reaches full funding level and will remain at that level indefinitely. This increase in funding ratio does not necessarily mean that the plan's actual funding status is improving. This is best illustrated by the case of Vermont State Teachers

* Because of the calculation of the initial unfunded liability, this method sometimes is also called "frozen initial liability."

Retirement System (VSTRS). Prior to 2006, VSTRS used frozen initial liability as the actuarial cost method. The initial unfunded liability was frozen in 1988 and is being paid off on a 30-year schedule (State of Vermont, 2005). Since no unfunded liability after 1988 would be recognized, the funding ratio has been improving gradually from 79.2 percent in 1994 to 90.7 percent in 2005, even through the severe stock market decline in the early 2000s (Buck Consultants, 2006). However, based on EAN cost method, the funding ratio of VSTRS actually decreased gradually from about 92 percent in 2001 to about 81 percent in 2005 and the Commission set up to study VSTRS recommended switching the cost method to EAN (State of Vermont, 2005). VSTRS adopted EAN cost method and the funding ratio rebounded to 84.6 percent in 2006 (Buck Consultants, 2006).

3.3 Asset Valuation

A discussion of actuarial valuation of pension plans is not complete without a discussion of the asset valuation method. While an actuarial cost method determines the liability side of the pension plan and how the cost should be allocated, the asset valuation method determines how the assets of a pension plan should be valued.

The easiest and simplest method to measure asset value is to use the fair market value, which is the fair value at which the asset is expected to be sold at the time of valuation. While fair market value provides an accurate estimation of the value of pension assets, it is not the asset value usually used in actuarial valuation. Of all the assumptions built into actuarial valuation, the assumption that is most difficult to predict, at least in the short run, is the investment return. The assumed rate of return is an average return in the long term and is unlikely to be exactly the same as the yearly investment returns. If the short-term return is higher than the assumed return, the pension plan can be overfunded and a funding surplus is created. If the short-term return is lower, a UAAL can occur. As the periodic pension contribution consists of both normal cost and amortization of UAAL (whether positive or negative), the volatility in funding ratio as a result of annual investment return can lead to volatility in pension contribution.

To reduce such volatility and to better reflect the long-term nature of pension investment, asset valuation is based on the actuarial value rather than the simple fair market value, although the former is related to the latter. A smoothing technique is used to arrive at an actuarial value of assets (AVA). The technique smoothes annual investment returns over a period of time, typically three to five years. If the smoothing period is five years, then only 20 percent of each year's return (whether positive or negative) is factored into that year's asset valuation. The remaining 80 percent will be gradually recognized in the actuarial value over the next four years. In other words, each year's actuarial value is based on 20 percent of the investment return of that year plus 20 percent of the return for each of the past four years. This averaged annual return over a multiyear period will also be closer to the assumed

Table 3.4 Arizona State Retirement System Assets (in $Millions)

Year	*Actuarial Value of Assets ($)*	*Net Assets ($)*
1998	15,577	19,930
1999	18,043	22,427
2000	20,292	23,926
2001	21,888	21,731
2002	22,642	19,210
2003	22,572	18,730
2004	22,659	21,431
2005	22,808	22,607

Source: Arizona State Retirement System CAFR, various years.

long-term rate of return than any single annual rate of return. In this way, any substantial increase or decrease in the fair value of assets will be recognized gradually, resulting in a much steadier and smoother actuarial value of assets compared to the fair market value of pension assets. This, in turn, leads to a steadier pension contribution for the plan sponsor.

How will AVA differ from fair market value under different financial market conditions? When the stock market is on the rise, generally the AVA should be expected to be lower than the fair market value of assets because part of the gain in value will not be recognized for several more years. If the stock market is in decline, the AVA should be expected to be higher than the fair market value as part of the losses will not be factored in for several more years. Table 3.4 compares Arizona State Retirement System's net asset value and the AVA between 1998 and 2005. From 1998 to 2000, when the stock market boomed, the AVA was substantially lower than the fair value of net assets. However, in the following three years when the stock market suffered heavy losses, the AVA was higher than the fair value of net assets by a substantial margin. It was not until 2004 and 2005 when the two values grew closer. It is obvious from this table that while the market value of net assets experienced considerable volatility over this period, the actuarial value of assets grew at a very steady pace and suffered only a minor decrease in 2003.

3.4 Financial Reporting and Accounting Standards

Understanding actuarial valuation is critical to understanding financial accounting and reporting related to pension plans. Since its creation in 1984, Governmental Accounting Standards Board (GASB) has been working on the standards of financial accounting and reporting for public pension plans and their plan sponsors. This work began with GASB Statement 1 in 1984, *Authoritative Status of NCGA*

Pronouncements and AICPA Industry Audit Guide. It continued in GASB Statement 5 in 1986, *Disclosure of Pension Information by Public Employee Retirement Systems and State and Local Governmental Employers*. It culminated in two statements in 1994, GASB Statement 25, *Financial Reporting for Defined Benefit Pension Plans and Note Disclosure for Defined Contribution Plans*, and GASB Statement 27, *Accounting for Pensions by State and Local Governmental Employers*. What follows is a discussion of the main standards incorporated in GASB 25 and 27.

3.4.1 GASB 25

GASB 25 establishes the standards of financial accounting and reporting for public pension plans. It consists of two main parts: (1) the financial reporting framework in terms of what information should be included in a financial report and (2) parameters for actuarial valuation.

3.4.1.1 Reporting Framework

GASB 25 distinguishes between two types of information: (1) current financial information about plan assets and financial activities and (2) actuarially determined information, from a long-term perspective, about the funded status of the plan. To disclose these two types of information, the reporting framework required by GASB 25 includes two sets of reports. In the first set, there are two basic financial statements that disclose current financial information: *statement of plan net assets* and *statement of changes in plan net assets*, followed by notes to the statements. In the second set of reports, there are two required schedules that disclose actuarially determined information: *schedule of funding progress* and *schedule of employer contributions*, also to be followed by notes to schedules.*

Two basic financial statements and notes to the statements:

1. A statement of plan net assets includes information about the plan assets, liabilities, and net assets as of the end of the plan's fiscal year. This statement provides information on the fair market value and composition of net assets. Table 3.5 presents a relatively simple example of this statement. It is the statement of net assets for the State Employees' Plan within the Delaware Public Employees' Retirement System.
   The statement of net assets contains three components: assets, liabilities, and net assets. Assets are listed in the order of liquidity: cash, receivables, and investments. As the statement is prepared on an accrual basis, any economic

* GASB 25, Paragraph 19.

Table 3.5 Statement of Plan Net Assets as of June 30, 2006 (in $Thousands)

Assets	**$**
Cash	1,638
Receivables	
Employer contributions	4,003
Member contributions	1,885
Total receivables	5,888
Investments, at fair value	
Domestic fixed income	1,035,969
Domestic equities	1,608,808
Pooled equity and fixed income	1,901,849
Alternative investments	729,249
Short term and money markets	71,842
Foreign equities	779,020
Total investments	6,126,737
Total assets	6,134,263
Liabilities	
Benefits payable	890
Accrued administrative expenses	229
Total Liabilities	1,119
Net assets held in trust for pension benefits	**6,133,144**

Source: Delaware Public Employees' Retirement System CAFR for fiscal year 2006.

resources, such as employer and employee pension contribution, earned but not yet received before the end of the fiscal year will be recorded as Accounts Receivable. Investments, always the most important part of assets for pension plans, are valued using fair market value of investment securities, such as bonds and stocks. The fair market value of each component of the investments also tells us something about the asset allocation of the pension plan (a subject that will be covered in Chapter 5). After subtracting some short-term liabilities from the total assets, the system's net assets are arrived at. Net assets are the source of funds available to pay for promised pension benefits in the future. What is missing from the statement of net assets is information on long-term pension liability and whether the net assets can cover the pension liability already accumulated. Such information is only available in the schedule of funding progress.

Table 3.6 Statement of Changes in the Plan Net Assets (in $Thousands)

Additions	**$**
Contributions	
Employer contribution	91,013
Member contribution	41,139
Transfer of contributions from PRI Fund	38,306
Total contributions	170,458
Investment	
Net increase in fair value	602,359
Investment earnings	82,638
Investment income	684,997
Less investment manager/advisor/custody fees	(17,977)
Less investment administrative expenses	(352)
Total net investment activity income	666,668
Total additions	837,126
Deductions	
Benefit payments	300,321
Refunds of contributions to members	2,880
Group life payments	4,779
Administrative expenses	4,871
Allocation of administrative expenses	(3306)
Total deductions	312,491
Net Increase	**524,635**
Net assets held in trust for pension benefits	
Beginning of year	5,608,509
End of year	6,133,144

Source: Delaware Public Employees' Retirement System CAFR for fiscal year 2006.

2. A statement of changes in plan net assets includes information about the additions to, deductions from, and net increase (or decrease) for the year in plan net assets. As seen in Table 3.6, the net assets for the Delaware State Employees' Plan increased by nearly $525 million in FY 2006, an almost 10 percent increase in net assets from the previous year. This increase in net assets primarily resulted from the investment activities. The fair value of investments increased by $602 million, largely due to the strong return in the stock market in that year. The other major addition to net assets was pension contributions from both employer and employees. Pension contributions,

however, were more than offset by pension benefit payments in that year plus refunds of contributions to plan participants who terminated employment in that year.

3. In the notes to the two financial statements, a brief description of the plan needs to be provided. The notes include information on what type of plan it is, i.e., single or agent multiple-employer plan, types of employees covered, benefits provision, accounting policies and funding policy as to how the pension contribution is determined. In this case, the State Employees' Plan is a cost-sharing, multiple employer plan and is funded on an actuarial basis.

Two required schedules and notes to the schedules:

1. A required schedule of funding progress includes historical information about the actuarially determined funded status of the plan from a long-term, ongoing plan perspective and the progress made in accumulating sufficient assets to pay benefits when due.* It reports the actuarial value of assets, the actuarial accrued liability, and the funding ratio, as shown in Table 3.7 on the funding progress of the Delaware State Employees' Plan.
   The funding ratio of the Delaware State Employees' Plan, while decreasing from 117 percent to just over 100 percent, has been consistently overfunded over this period, an achievement matched by very few state-level pension plans. Therefore, the plan has a surplus, which is shown as a negative UAAL. A measurement of the relative size of UAAL (whether positive or negative) is the ratio of UAAL to covered payroll. In case of a positive UAAL, it is similar to measuring a nation's debt as a percentage of its overall economy. A decreasing ratio indicates that it is becoming easier to service the debt even if the absolute debt size may still be increasing.
2. A required schedule of employer contributions includes historical trend information about the annual required contributions of the employer(s) (ARC) and the contributions made by the employer(s) in relation to the ARC. GASB 25 requires at least six consecutive years of historical information for both schedules.† Table 3.8 shows this schedule for the Delaware state plan, which fully paid its required pension contributions over the entire period.
   Because of their long-term perspective, these two schedules are important in assessing the financial health of a pension fund and the extent to which a government is meeting its financial obligation. Therefore, they garner more attention than the two statements.

* GASB 25, Paragraph 19.c.

†GASB 25, Paragraphs 37 and 38.

Table 3.7 Schedule of Funding Progress (in $Million)

	(a)	*(b)*			*(c)*	
Actuarial Valuation Date	*Actuarial Value of Assets*	*Actuarial Accrued Liability*	*Unfunded Actuarial Accrued Liability (b-a)*	*Funded Ratio (a/b)*	*Covered Payroll*	*UAAL as a Percentage of Covered Payroll ((b-a)/c)*
6/30/1999	3,888	3,350	(538)	116.1	1,125	-48
6/30/2000	4,409	3,769	(639)	117.0	1,198	-53
6/30/2001	4,759	4,232	(526)	112.4	1,270	-42
6/30/2002	4,956	4,521	(434)	109.6	1,351	-32
6/30/2003	5,125	4,794	(330)	106.9	1,355	-24
6/30/2004	5,387	5,229	(157)	103.0	1,399	-11
6/30/2005	5,660	5,572	(887)	101.6	1,474	-6
6/30/2006	5,998	5,901	(97)	101.7	1,589	-6

Source: Delaware Public Employees' Retirement System CAFR for fiscal year 2006.

3.4.1.2 Parameters

To help the users of financial reports understand how the actuarial values are derived, GASB 25 also requires the plan to publish a note to the schedule, which discloses certain parameters of actuarial valuation as well as any factors that affect the trend, such as change of actuarial cost method. GASB 25 sets the standards for parameters on the actuarial valuation. Following is a list of some of the important parameters.

1. **Economic assumptions**: The investment-return assumption should be based on an estimated long-term investment yield. The major economic assumptions underlying the actuarial valuation, such as the investment return assumption and projected salary increase assumption, should include the same assumption with respect to inflation.*
2. **Actuarial cost methods**: As mentioned earlier in this chapter, GASB recommends that one of six actuarial cost methods should be used.† Through actuarial studies, GASB found that over a long period of time, each of these methods yields funding and asset accumulation that are sufficiently similar to each other.‡

---

* GASB 25, paragraph 36c.
† GASB 25, paragraph 36d.
‡ GASB 25, paragraph 141.

Table 3.8 Schedule of Employer Contribution (in $Thousands)

For the Year Ended June 30	*Annual Required Contributions*	*Percentage Contributed*
1999	52,338	100
2000	43,433	100
2001	20,542	100
2002	18,148	100
2003	40,175	100
2004	57,459	100
2005	70,638	100
2006	91,013	100

Source: Delaware Public Employees' Retirement System CAFR for fiscal year 2006.

3. **Asset valuation**: The actuarial value of asset should be market related, meaning it can be either market value or a value that recognizes changes in market value over a period of time.*
4. **Amortization**: The amortization of UAAL should meet the following requirements: †

    1. The maximum amortization period was 40 years within the first 10 years of the publication of GASB 25 and 30 years afterward.
    2. The amortization period can be both an open and a closed period. For a closed period, if the amortization period is set at 30 years initially, with each passing year, the amortization period will decrease by one year until it reaches zero in year 30. For an open period, the amortization period remains at 30 years, meaning the remaining UAAL will be reamortized over a 30-year period every year. The use of an open approach helps reduce volatility in the employer's pension contribution rates. GASB finds both approaches to be acceptable from a long-term, ongoing plan perspective.‡
    3. The total unfunded liability can be amortized in both level dollar and level percentage of the projected payroll of active plan members.§

In the case of Delaware State Employees' Plan, the parameters for the actuarial valuation of the pension disclosed in its 2006 comprehensive annual financial report (CAFR) were:

* GASB 25, paragraph 36e.
† GASB 25, paragraph 36f.
‡ GASB 25, paragraphs 145 and 146.
§ Ibid.

- The actuarial method is EAN
- Amortization method is level percentage with a closed period for plan base and open period for aggregate plan gains/losses
- Asset valuation method is a five-year smoothed market value
- Investment rate of return is 8 percent
- Cost-of-living adjustments are ad hoc

The disclosure of parameters is particularly important when changes in certain key parameters can have a major effect on the funding situation of a pension plan. For example, the funding ratio of Virginia Retirement System (2006) dropped from 90.3 percent in 2004 to 81.3 percent in 2005, a considerable drop given the fact of the funding ratio of most other large pension plans stabilized in 2005. One reason for this drop is that VRS changed its assumed rate of return from 8 percent in 2004 to 7.5 percent in 2005, thus causing an increase in its accrued actuarial liability and a drop in its funding ratio. While this drop in funding ratio is of concern, it also increases the chance that VRS will be able to meet its financial goals in the future.

3.4.2 GASB 27

This statement establishes standards of accounting and financial reporting for pension expenditures/expense and related pension plan information in the financial reports of state and local government employers. The objective of this statement is that the measurement of employer's pension expenditures/expense for an accounting period in the employer's financial report should be consistent with the employer's required contributions for that period and related information reported by the pension plan.

This consistency entails the calculation of the employer's annual pension cost and net pension obligation (NPO) by using the same actuarial methods and assumptions that are applied in determining the plan's funding requirements, as discussed in GASB 25. That means the employer's annual pension cost reported in its financial report should be equal to the annual required contributions of the employer (ARC) to the plan for that year. However, when the employer contributes less than the ARC in one year, the contribution deficiency constitutes a net pension obligation and the employer's annual pension cost in the subsequent year will no longer be equal to its ARC.

When a contribution deficiency occurs, the employer's annual pension cost, prepared for accounting purposes, is equal to the ARC, plus one year's interest on the NPO, and an adjustment to the ARC.* The adjustment is needed due to the fact that the ARC, prepared for funding purposes, already includes an amount for

* GASB 27, Paragraph 12.

the amortization of the deficiency. Because the entire deficiency is already recognized as pension cost in the prior year, the amortization amount of the deficiency in the ARC should not be recognized again for accounting purposes in the next year. It, therefore, should be subtracted from the ARC to calculate the employer's annual pension cost. The interest on the NPO is added to the employer's annual pension cost to account for the loss of interest on the contribution deficiency to the pension plan. The net result of the adjustment should be that annual pension cost is approximately equal to the ARC if the employer didn't have an NPO. The calculation of the adjustment is based on the same amortization methodology used in determining the ARC for that year and the same actuarial assumptions.* This adjustment has to be recalculated for each year. While an NPO is created due to contribution deficiency and, thus. is a liability to the employer, it can also occur due to contribution excess. The NPO, in this case, is negative and an asset to the employer. Calculation of the annual pension cost is just the opposite. The one-year interest on the negative NPO is added to and the adjustment is subtracted from the ARC to arrive at the annual pension cost.† The NPO at the end of the period is equal to NPO at the beginning of the period, plus the annual pension cost for the period and minus the actual contribution made.

Net pension obligation indicates the extent to which the employer has met its financial obligation in a particular year as well as overtime. While an absence of NPO and a decreasing NPO indicates a full commitment to the pension obligation, a gradually increasing NPO is a sign that the government employer is under increasing financial stress. Table 3.9 presents an example of an increasing NPO. The Teachers' Retirement System (TRS) is one of several state retirement systems sponsored by the Illinois state government. This table shows that the Illinois state government has cumulatively underpaid its pension contribution to the TRS by

Table 3.9 Illinois State Net Pension Obligation for Teachers' Retirement System (in $Thousands)

Actuarially required contribution (ARC)	1,679,524
Plus: Interest on net pension obligation (NPO)	610,630
Adjustment to the ARC	(364,011)
Annual pension cost	1,926,143
Employer contribution	601,472
Increase in NPO	1,324,671
NPO at June 30, 2005	7,183,894
NPO at June 30, 2006	8,508,555

Source: Illinois State CAFR for fiscal year ended June 30, 2006.

* GASB 27, Paragraph 13.

† Ibid.

$8.5 billion as of 2006 and, more importantly, it shows that the NPO increased by about 20 percent in 2006.

3.4.3 GASB Statement 50

In May 2007, GASB issued Statement 50, *Pension Disclosures—An amendment of GASB Statements No. 25 and No. 27.* This statement is targeted at pension plans using the aggregate cost method in actuarial valuation. As discussed earlier, for pension plans using the aggregate cost actuarial valuation method, no UAAL is separately identified and, therefore, no funding schedule is necessary and available under GASB 25 and 27. The funding ratio of these plans will always be equal to 100 percent. GASB only requires the plans that use the aggregate actuarial cost method to disclose that the method does not identify or separately amortize unfunded actuarial liabilities.* Since the lack of funding schedule makes it impossible to see if the pension plan is making any progress toward full funding, GASB amended GASB 25 and 27 by requiring plans using the aggregate cost method to separately calculate its accrued actuarial liability (AAL) by using the EAN cost method. With AAL known, the plan can calculate its funding ratio and publish a funding schedule. This also makes it possible to compare such plans with other public plans. This amendment to GASB 25 and 27 is meant to bring more transparency and accountability to pension plans that use the aggregate cost method.

References

Bizley, M.T.L. 1950. The Effect on Pension Fund Contributions of a Change in the Rate of Interest. *Journal of the Institute of Actuaries Students' Society* 10: 47-51.

Buck Consultants. 2006. *Report on the Fifty-Ninth Actuarial Valuation of the State Teachers' Retirement System of Vermont.* Prepared as of June 30, 2006. http://www.vermonttreasurer.gov/documents/retireTeacher/reports/teacherActuary2006.pdf (Accessed May 2, 2007.)

State of Vermont, Commission on Funding the Vermont State Teachers' Retirement System. 2005. *Report of the Commission on Funding the Vermont State Teachers' Retirement System.* http://www.vermonttreasurer.gov/documents/retireTeacher/reports/teacherCommissionFinalReport200511.pdf (Accessed May 2, 2007.)

Virginia Retirement System. 2006. *Comprehensive Annual Financial Report* for the year ended June 30, 2006. http://www.varetire.org/Pdf/Publications/2006annurept.pdf (Assessed May 4, 2007.)

Wisconsin Legislative Council. 2005. *2004 Comparative Study of Major Public Employee Retirement Systems.* www.legis.state.wi.us/lc/publications/crs/2004_retirement.pdf (Accessed June 6, 2006.)

* GASB 25, paragraph 40.

Chapter 4

Governing Public Pension Plans

It is almost impossible to discuss public pension plan governance without first discussing the difference in the legal framework governing public and private pension plans. The difference is that private pension plans are governed by the federal Employee Retirement Income Security Act (ERISA) of 1974, whereas public pension plans are exempt from ERISA.

4.1 ERISA

ERISA is the most comprehensive federal regulation of private pension plans. It was the culmination of a long list of federal legislation on employee benefit plans. The closing of the Studebaker plant in South Bend, Indiana, in 1964, which inflicted heavy pension losses on workers, led to congressional hearings on pension benefits, and these hearings eventually led to the passage of the Employee Retirement Income Security Act (ERISA) on Labor Day in September 1974 (General Accounting Office, 1979).

ERISA does not require any private employer to establish a pension plan. It only requires that those who establish plans must meet certain minimum standards. ERISA prescribes standards for plan participation, vesting, funding, fiduciary duties, disclosure, and reporting. It also provides mechanisms to enforce these standards and to ensure that employees receive some of their accrued pension benefits. The main goal of ERISA is to prevent abuses of private pension plans and protect the benefits of pension beneficiaries. These standards and rules are contained in four titles.

4.1.1 ERISA Standards

Title I establishes minimum requirement for participation, coverage, vesting, funding, fiduciary standards, and reporting and disclosure. Title I is enforced by the Employee Benefits Security Administration of the U.S. Department of Labor.

Part 1 of Title I requires the administrator of an employee benefit plan to furnish participants and beneficiaries with a summary plan description, their rights, benefits, and responsibilities under the plan. He is also required to furnish participants with a summary of any material changes to the plan. The administrator must file an annual report each year with the Department of Labor, containing financial and other information concerning the operation of the plan, and the report must be audited by an independent public accountant. The summary of the information in the annual report must also be given to plan participants and beneficiaries.

Part 2 of Title I sets the minimum standard for participation and vesting. The minimum standard for participation in a pension plan is that an employee either attains the age of 21 or completes one year of service. Prior to ERISA, 53 percent of workers in plans with vesting provision needed to work for 15 years before they could qualify for full vesting.* ERISA sets a maximum of 10 years for full vesting. This part also contains an anticutback rule, which, with narrow exceptions, does not allow a pension plan to decrease the accrued benefit of a participant through an amendment of the plan.

Part 3 of Title I sets the minimum funding standards. Employers are required to fund the normal cost plus an amount to amortize the unfunded accrued liability of a plan. The maximum amortization period is 40 years for plans established before 1974 and 30 years for plans established after 1974.

Part 4 of Title I sets the standards and rules governing the conduct of plan fiduciaries. ERISA considers a person as fiduciary of a pension plan if (1) he exercises any discretionary authority or discretionary control respecting management of such plan or exercises any authority or control respecting management or disposition of its assets; (2) he renders investment advice for a fee or other compensation, direct or indirect, with respect to any moneys or other property of such plan, or has any authority or responsibility to do so; or (3) he has any discretionary authority in the administration of such plan.† Fiduciaries are required to discharge their duties "solely in the interest of plan participants and beneficiaries and for the exclusive

---

* U.S. Department of Labor, Bureau of Labor Statistics. Defined Benefit Plans at the Dawn of ERISA. http://www.bls.gov/opub/cwc/cm20050325ar01p1.htm (Accessed 7/27/06.)

† § 3(21)(A) Fiduciary § 1002(21)(A)

purpose of providing benefits and defraying reasonable expenses of administering the plan." The fiduciary must discharge such duties

> with the care, skill, prudence, and diligence under the circumstances then prevailing that a prudent man acting in a like capacity and familiar with such matters would use in the conduct of an enterprise of a like character and with like aims; by diversifying the investments of the plan so as to minimize the risk of large losses, unless under the circumstances it is clearly present not to do so; and in accordance with the documents and instruments governing the plan insofar as such documents and instruments are consistent with the provisions of this title and Title IV.*

This standard set by ERISA for fiduciary responsibilities is commonly known as the "prudent expert" rule.

ERISA also prohibits the fiduciary from engaging in certain transactions. A fiduciary shall not (1) deal with the assets of the plan in his own interest or for his own account, (2) in his individual or in any other capacity act in any transaction involving the plan on behalf of a party (or represent a party) whose interests are adverse to the interests of the plan or the interests of its participants or beneficiaries, or (3) receive any consideration for his own personal account from any party dealing with such plan in connection with a transaction involving the assets of the plan. This standard is commonly known as the conflict of interest rule or code of ethics. ERISA also holds a fiduciary personally liable for breaches of any of the responsibilities, obligations, or duties imposed upon fiduciaries by this title:

> Any person who is a fiduciary with respect to a plan who breaches any of the responsibilities, obligations, or duties imposed upon fiduciaries by this subchapter shall be personally liable to make good to such plan any losses to the plan resulting from each such breach, and to restore to such plan any profits of such fiduciary which have been made through use of assets of the plan by the fiduciary, and shall be subject to such other equitable or remedial relief as the court may deem appropriate, including removal of such fiduciary.

Title II of ERISA contains standards that must be met by employee pension benefit plans in order to qualify for favorable tax treatment. Noncompliance with these tax qualification requirements of ERISA may result in disqualification of a plan and/or other penalties. The Internal Revenue Service (IRS) of the Department of Treasury administers Title II of ERISA. Title III contains provisions regarding administration and enforcement of the ERISA requirements. Title IV establishes

* § 404 Fiduciary duties (§ 1104)

the private pension insurance program by creating the Pension Benefit Guaranty Corporation (PBGC), a government entity that insures protection for defined benefit plans that terminate without sufficient assets. The assets used to guarantee the pension benefits come from the insurance premium paid by the employers and the pension assets transferred to the PBGC once the employer terminates the pension plan. There is, however, a cap on the benefit payout from the PBGC.

In all, ERISA sets in place a legal wall of protection for pension benefits for private sector employees by requiring plan administrators to report and disclose plan information regularly, by setting minimum funding standards, by requiring plan fiduciaries to manage plan assets responsibly, and by insuring plan participants' and beneficiaries' pension benefits.

4.1.2 ERISA and Public Pension Plans

In general, ERISA does not cover plans established or maintained by state and local governments. At the time of passage in 1974, Congress excluded governmental retirement systems from the major provisions of ERISA pending further study of the need for federal regulation of governmental plans. In March 1978, in accordance with the ERISA mandate, the House of Representatives Committee on Education and Labor (1978) issued a report to the Congress. The House Committee's Pension Task Force estimated that about 42 percent of the defined benefit plans in the public sector were funded in ways not related to their accrued pension liabilities, either using the pay-as-you-go method or some other nonactuarial method, such as matching of employee contributions. Adopting a funding standard similar to that required by ERISA would require many of these governments to raise their contributions by more than 100 percent, and a few by more than 400 percent. In that year, a bill called the Public Employee Retirement Income Security Act of 1978 was introduced in the House to regulate public pension plans. It, however, failed to pass.

4.2 Public Pension Benefit Protection

Despite the lack of federal regulation, state and local governments over the years have established a body of laws and regulations that in aggregate are substantially similar to ERISA in terms of the protection of plan participants' benefits, vesting requirement, financial reporting, and fiduciary responsibility standards. In this section, we examine the legal protection of pension benefits and, in the next two sections, we examine the administration and oversight of public pension plans. Whenever possible, the practice in the public sector is compared to the standards of ERISA.

Even though there is no pension benefit insurance program in the public sector, employees in the public sector enjoy a higher level of protection of their pension benefits than their counterparts in the private sector. There are two reasons for this stronger protection in the public sector. First and foremost, the nature of the employer who establishes the pension plan and the pledge to pay for pension benefits is different. The difference between private companies and government entities is that the former can file for bankruptcy and eventually liquidate. When that happens, the assets in the private pension plans will be transferred to the PBGC who will then pay for the accrued benefits of the plan participants up to the limit set by the PBGC, and the participant will have lost any opportunity to accrue benefits with the same employer in the future. As discussed in Chapter 2, since much of the defined benefit accrues later in one's career, this involuntary termination can lead to substantially smaller accumulation of pension benefits. In the public sector, state and local government entities are bankruptcy remote. While a handful of municipalities filed for bankruptcy in the past, municipal bankruptcy is very different from corporate bankruptcy. It does not lead to liquidation of the municipality, but it still has to pay all its financial obligations after coming out of bankruptcy. Therefore, public sector employees do not have to fear losing accrued benefits due to the bankruptcy of the government employer who sponsors the pension plan.

The nature of the pledge to pay pension benefits is also different in the public sector. While theoretically it is the assets in the pension plan, which are used to pay for pension benefits, government sponsors of pension plans are ultimately responsible for paying the pension benefits promised to government employees. In other words, it is the government entity's ability to collect revenue that is the ultimate security behind the payment of pension benefits. As long as the revenue base does not completely erode, the government employer will need to find revenue to pay for the promised benefits even if there are not sufficient assets in the pension plan to do so.

This assurance of having the financial resources to pay for pension benefits does not amount to much if the pension benefits themselves are not protected. The second major reason for stronger pension benefit protection in the public sector is that this protection covers not only accrued benefits, but also benefits yet to be accrued. Even though ERISA protects against any reduction in the accrued benefits, it offers no protection against reduction in future pension benefits yet to be earned. This means private pension plan sponsors can change the pension plan at any time during the period an employee is working at the company. The company can change the pension benefits formula, or it can simply stop the current plan and start a completely different plan. If such change leads to a reduction in future pension benefits accrual compared to that under the old plan, there are no specific federal laws that plan participants can use to prevent the company from doing so.

In the public sector, plan participants have legal protection against reduction in not only accrued benefits, but also promised future benefits that are yet to be earned as long as the plan participant stays with the same government employer.

In other words, it is the promise of pension benefits at the time of initial participation in the benefit program that is protected. The stronger legal protection against impairment of future pension benefits is derived from state constitutions, statutes, and case laws. Embedded in such protection is the concept of "contractual right." Once a person enters into employment with a government entity and starts earning pension benefits, he thus earns the "contractual rights" to all future pension benefits as long as he is vested and continues to work for the same government employer. Such "contractual rights" are protected by the U.S. Constitution and state constitutions.

According to a survey by the National Council on Teacher Retirement, nine states have constitutional guarantee of public pension rights and another twenty states have statutory guarantee of such benefits (Moore, 2005). An example of constitutional guarantee can be found in Illinois. Article XIII, section 5, of the Illinois Constitution, which pertains to pension and retirement rights, provides that:

> Membership in any pension or retirement system of the State, any unit of local government or school district, or any agency or instrumentality thereof, shall be an enforceable contractual relationship, the benefits of which shall not be diminished or impaired.

As for an example of statutory protection of pension benefit, Section 692 of Kentucky State Statute 61, *Benefits not to be reduced or impaired—Exception*, stipulates:

> It is hereby declared that in consideration of the contributions by the members and in further consideration of benefits received by the state from the member's employment, KRS 61.510 to 61.705 shall, except as provided in KRS 6.696 effective September 16, 1993, constitute an inviolable contract of the Commonwealth, and the benefits provided therein shall, except as provided in KRS 6.696, not be subject to reduction or impairment by alteration, amendment, or repeal.

For states with or without constitutional or statutory protection, court decisions have also established protection of pension benefit. Such court decisions, usually in favor of plan participants and beneficiaries, are based either on the specific state constitutional guarantee of pension benefit or on the more general contractual right guaranteed by U.S. and state constitutions. For example, in *Felt v. Board of Trustees of the Judges Retirement System*, Illinois Supreme Court found unconstitutional an amendment to the Illinois Pension Code that changed the salary base for determining pension benefits from the judge's salary on the final day of service to the average salary over the last year in service.* The Court found that the amendment violated

* 107 Ill.2d 158, 89 Ill.Dec. 855, 481 N.E.2d 698 (1985)

the constitutional right of judges because it diminished their retirement benefits and impaired their contract rights under the Illinois Constitution.

In some cases, the protection provided by constitutional guarantee and contractual right is extended beyond the pension benefit itself to include protection of financial resources, such as government employer pension contribution, used to pay for pension benefit:

- In the early 1990s, the New York state government skipped pension contribution to the state pension system as a result of fiscal stress. The court ruled in the subsequent lawsuit that such action was unconstitutional because it impaired the pension benefit of the pension system's members and beneficiaries.*
- In the late 1990s, Hawaii state and county governments underfunded pension contribution by about $350 million. State of Hawaii Organization of Police Officers, later joined by the trustees of the state retirement system, sued the state. In 2007, the Hawaiian Supreme Court ruled that the state's action was unconstitutional as it violates the nonimpairment clause of accrued benefits in the state constitution.† Although the Supreme Court did not require the state government to repay the retirement system, the ruling effectively prohibits the state government from underfunding pension contribution again in the future.
- In 2003, facing massive budget deficit, the California state government withheld $500 million in pension payments to California State Teachers Retirement System (CalSTRS). CalSTRS sued the state government. In 2005, a Sacramento Superior Court ruled that the state government's action violated state constitution. California finally paid $500 million to CalSTRS in 2007.

Because of such strong legal protection, there appears in the public sector a unique phenomenon of the tiered pension system. When the plan sponsors find that the pension benefits are too expensive and not affordable in the future, they cannot make changes to the pension benefits of participants already in the plan. However, they can establish a new pension plan with reduced pension benefits for future new employees who have not earned any "contractual rights" to the old pension benefits. Thus, a two-tiered pension system is created. For example, New York State has a four-tiered pension system, due to new pension legislation enacted in 1973, 1976, and 1983 that divided the workforce into four tiers. In recent years, several states established either mandatory or optional defined contribution pension plans. In both cases, current participants in defined benefit plans are given

* More discussion of the New York State court decision can be found in the case study on New York State pension plan management in Chapter 6.

† *Georgia Kaho'Ohanohano, et al. vs. State of Hawaii.* The full text of the Supreme Court ruling can be read at http://www.state.hi.us/jud/opinions/sct/2007/26178.pdf (Accessed September 19, 2007.)

the option of staying in the old defined benefit plan or joining the new defined contribution plan. In the private sector, when a company switches to a defined contribution plan from a defined benefit plan, all employers, new and old, will have to enroll in the new defined contribution plan.

For the minimum vesting requirement prescribed in ERISA to protect employees, public pension plans have substantially similar standards. As explained in Chapter 2, no public pension plans have a vesting period longer than 10 years, the maximum allowed by ERISA, and most plans adopt a five-year vesting period. The average vesting period has also been decreased considerably over the past 25 years.

4.3 Public Pension Plan Administration

Even without federal regulation of pension administration, state and local governments have adopted legislations that are also similar to ERISA in many aspects of pension plan administration. In this and the next section, we examine the major aspects of public pension plan administration: administrative structure, Board and staff responsibilities, internal control, financial reporting, funding policy, and oversight.

As briefly explained in Chapter 1, public pension plans are administered by public employee retirement systems (PERS), legal entities specifically set up by the sponsors of pension plans to administer such plans. A PERS can manage either one or multiple public pension plans.

4.3.1 Pension Plan Administration: The Governing Board

With few exceptions, the typical administrative structure of a retirement system consists of a board of trustees and a supporting staff headed by an executive director.* As the board of trustees establishes the overall policy for the operation of the pension system, it plays the most critical role in pension plan administration.

* Of the major state level retirement systems, those in Florida, Iowa, and Washington do not have an independent governing board. In Florida, the retirement system is managed by the Division of Retirement within the Department of Management Services. The governor appoints the department's secretary who appoints the director of the division. In Iowa, the retirement system is an independent agency within the executive branch of the government. In Washington, the Department of Retirement Systems manages several state-level retirement systems. The department's director is appointed by the governor. However, in all of these cases, the most important management activity of a retirement system, namely the pension asset investment management, lies with a separate independent investment board.

4.3.1.1 Election of Trustees to the Board

In order to be a trustee of a governing board, the person has to be either elected by plan participants, appointed by the plan sponsor, or serve as *ex officio*. Trustees may be elected by either active members or retired plan members, and they themselves may be active or retired plan members. Appointments are typically made by a chief elected official, such as the governor or mayor, or by a governing body, such as a state or local legislative body. Some trustees serve on the board by virtue of their holding a particular public office, such as that of state treasurer or controller.

The governing boards vary significantly in terms of the number of trustees and the board composition. According to a survey of 86 large state and local public pension boards by the National Education Association (2006), the number of trustees varies from one in New York to 26 for the University of California Retirement Plan, with the median being 9. About half of all trustees are either active members or retirees, and about 40 percent of the systems have a majority comprised of active and/or retired members. The selection method also shows significant variation. In some states, such as Arizona, the governor makes all the appointments to the governing board, whereas in other states, such as Arkansas, all appointments are made by the plan members and/or the legislative body. Some states require members of the governing board to have certain specific skills, especially skills in investment management. For example, Arizona requires that four trustees are not members of Arizona State Retirement System and have at least 10 years of substantial experience in investment, economics, or finance.

The New York state pension system has a unique board structure, with the board consisting of only one trustee. The state comptroller, an official elected statewide, is the sole trustee of the New York state pension system for state and local employees. As will be seen in Chapter 6, this unique governing structure plays a key role in the New York state public pension plan management.

4.3.1.2 Board's Fiduciary Responsibility Standard

Even though ERISA fiduciary rules do not apply to public pension plans, most state pension codes have languages with regard to fiduciary rules that are essentially the same as those in ERISA. Illinois is one typical example. In Section 1-109 of the Illinois Pension Code, fiduciary duties are defined as follows:

> A fiduciary with respect to a retirement system or pension fund established under this Code shall discharge his or her duties with respect to the retirement system or pension fund solely in the interest of the participants and beneficiaries and:
>
> (a) For the exclusive purpose of:
>
> (1) Providing benefits to participants and their beneficiaries; and

> (2) Defraying reasonable expenses of administering the retirement system or pension fund;
>
> (b) With the care, skill, prudence, and diligence under the circumstances then prevailing that a prudent man acting in a like capacity and familiar with such matters would use in the conduct of an enterprise of a like character with like aims;
>
> (c) By diversifying the investments of the retirement system or pension fund so as to minimize the risk of large losses, unless under the circumstances it is clearly prudent not to do so; and
>
> (d) In accordance with the provisions of the Article of the Pension Code governing the retirement system or pension fund.

By examining this language with that of ERISA discussed in the first section of this chapter, it is easy to find that they are almost identical.

4.3.1.3 Conflict of Interest Rule and Code of Ethics

Just like ERISA, trustees of public pension systems are also subject to conflict of interest rules and other ethics laws governing their behavior. According to the National Council on Teacher Retirement, 38 states have conflict of interest rules and all states have code of ethics laws (Moore, 2005). Conflict of interest clause and ethics laws are put in the board governance policy to prevent trustees from engaging in any activities and decision making that puts their own interest above that of the members they service. For trustees, the code covers such issues as acceptance of gratuities and what behavior constitutes conflict of interest. It provides guidance to trustees and instructs them to avoid certain practices that may adversely affect plan members.

Ohio Public Employees Retirement System (2005) board governance policy offers a typical example of such codes of conduct. In the conflict of interest rule, the policy states:

> Board members are prohibited by law from engaging in certain party-in-interest transactions (i.e., furnishing of goods or services between the system and a relative of a board member), and are prohibited from using assets of the system for their own interests. Board members are prohibited from receiving any consideration for their own personal account from any party dealing with the system in connection with a transaction involving the system assets. Board members may not act on behalf of a party whose interests are adverse to the system, its participants, or beneficiaries. The system is prohibited by law from making investments or doing business with individuals or entities controlled by individuals who were board members, officers, or employees of the

> system from being involved in investment recommendations to the system where such individuals or entities would benefit by any monetary gain. Board members are prohibited from having any direct or indirect interest in the gains or profits of any board investment.

In recognition of the importance of this policy, California Teachers' Retirement System (2006), the second largest public pension system in the country, adopted a more rigorous conflict of interest policy that went into effect in 2007: Full disclosure of communication initiated by a board member to a staff member or consultant if the communication could reasonably be interpreted as an attempt to influence a specified outcome regarding an investment transaction; 12-month recusal from any decision involving a campaign contributor or gift maker where the amount exceeds $250 to a board member.

As in the case of ERISA, state pension ethics policy also typically lays out the consequences when the code of ethics is breached by a trustee. For example, the Ohio pension ethics policy states that the failure of any board member or employee to abide by the Ethics policy will result in discipline, such as dismissal and potential civil or criminal sanctions. In 2006, a former Ohio State Teachers Retirement System (STRS) board member, Hazel Sidaway, was sentenced to two years probation and 200 hours of community service on convictions for two conflict of interest charges for accepting $670 worth of tickets to sporting and entertainment events from investment advisors to STRS (Ohio Ethics Commission, 5/12/2006). Four other STRS board members were also sentenced to one-year probation and 30 to 60 hours community service for Ethics Law conflict of interest violations for their acceptance of entertainment paid for by an investment advisor to STRS (9/19/2006).

In all, what is expected of the trustees of public pension plans, codified in state pension legislation and policies, is the same as that expected of the trustees of private pension plans, codified in ERISA.

4.3.1.4 The Board's Main Functions

The board's main functions in fulfilling its fiduciary responsibilities can be divided into two major categories: acquisition of sufficient assets to pay pension benefits and effective operation of the pension system.

Acquisition of sufficient assets — The board's fiduciary responsibility to the plan beneficiaries is to acquire the necessary assets for paying pension benefits. This is achieved through pension contribution from the plan sponsor and members and pension asset investment. For pension contribution, it requires the board to determine the pension contribution level under the advice of actuaries. Minimally, this involves choosing the actuarial valuation method and appropriate economic and demographic assumptions in calculating the pension liabilities and in determining the pension contribution rate that is sufficient to fund pension benefits. As will be discussed later in this chapter, not all plan governing boards have the full responsibility for setting the contribution rate. In some states, this rate is set by the plan sponsor itself through legislation. Even in plans where the governing board has the responsibility of setting the contribution rate, public pension boards typically do not force the plan sponsor to pay the annual required amount.

The governing board has more authority over the second aspect of asset acquisition, namely investment strategy. With the advice of experts, the governing board has the authority to design an investment policy with a particular focus on asset allocation. As pension asset investment management is by far the most important responsibility of the pension governing board and the pension system, this subject will be discussed in more detail in the next chapter.

To ensure the accumulation of sufficient assets, the governing board also needs to establish a risk management and control system to protect pension assets from loss, theft, or misuse, and major risks to the pension plan's long-term fiscal health can be identified.

Effective operation of the pension system — Given the complexity of public retirement systems, effective governance calls for the governing board to focus on policies with regard to the operation of the pension system and leave the day-to-day administration to staff through prudent delegation of authority. The board needs to establish clear roles and responsibilities for all key parties involved in the decision-making process, including the board, board committees, chief executive officer, and other key staff members, such as chief investment officer. Such clear expectation of rules and responsibilities is essential to the prudent delegation of authority. Along with the delegation, staff performance evaluation should also be conducted, based on established performance measures, to make sure their responsibilities are met.

4.3.2 Pension Plan Administration: The Staff

The responsibility of the staff of public pension system, under the leadership of the executive director, is to implement the policies designed by the governing board and manage the day-to-day administration of the pension system. Their duties can be divided into three main areas: member service, supporting service, and investment management.

4.3.2.1 Member Service

The staff provides a wide range of services to members and beneficiaries, from the time they first participate in the plan until the time they receive their last pension check:

1. Enrollment: The starting point of all member services is to enroll a new employee into the retirement system.
2. Consultation and education: The staff will help members project their retirement benefits for the members' financial planning purpose. If a member terminates employment, he can either leave his contributions in the system or withdraw his contributions. The staff can assist the member in determining the one that is more beneficial to the member. If the member wants to purchase service credit, the staff can assist in determining the cost. Prior to a member's retirement, the staff will also assist the member in calculating his or her annual benefits and application process. The pension system staff also holds periodic seminars on financial planning to help members plan for the future.
3. Communication: There are four main types of publications a pension system distributes to communicate with its members and retirees. First, there are the handbooks for members, acquainting them with the pension plan and all the benefits available to them as well as the rules for obtaining these benefits. Second, there are pamphlets published by the pension system that discuss some aspects of the pension plan and pension benefits in more detail than are discussed in the handbooks. Third, there are periodic newsletters published by the pension system that alert the members and retirees to the new developments in and changes to the pension plans. Fourth, the pension system also sends annual statements to members and retirees. The annual statement to the member contains information about her membership, beneficiary, service credits earned, and projected benefits. The statement to the retiree contains information about his/her annual retirement payment and tax withholding.
4. Loan program: Some pension systems also have loan programs for its members. The members can take out a loan against their own contributions, usually for financial emergencies, after becoming a member for a certain period of time. The repayment of the loan is made through payroll deduction and the loan has to be paid back with interest. For example, with the New York state retirement system, a member must have at least one year of member service credit to apply for a loan and he may borrow up to 75 percent of the contribution balance. The loan has to be repaid within five years with interest.
5. Disability: When a member becomes disabled and applies for disability benefits, the pension system decides whether the disability is permanent and disability benefits should be given. If the decision is not in favor of giving the member disability benefits, the member can appeal the decision and an administrative hearing ensues.
6. Retiree service: The staff determines each retiree's eligibility for retirement and his annual retirement payment. At the request of the retiree, the system can also directly deposit monthly payments into the retiree's bank account. If the retiree can no longer handle his finances due to incapacitation, the system will work with the person designated by the retiree to handle the

retiree's finances. If the retiree passes away, the deceased retiree's payments will stop and the pension system will start paying benefits to his designated beneficiary, if any.

While many of these services for members and retirees can be done through phone calls, mailings, and the Internet, some of these services require personal consultation. Therefore, a statewide pension system usually sets up service centers across the state to facilitate the service provision to the members and retirees, For example, in addition to its headquarters in the state capital, the New York state pension system has 15 service centers throughout the state.

4.3.2.2 Supporting Services

While providing direct services to the members and retirees is an important task performed by the retirement system staff, there are also many important supporting services performed by the administrative staff to make this possible:

1. Information system management: As a large pension system can contain tens of thousands and in many cases hundreds of thousands of members and retirees, providing services to them also involves information system management. Integrating advanced data processing technology into all aspects of retirement system management is quintessential in providing efficient and effective services to the members and retirees. This information management system is required to perform many important tasks and lies at the heart of pension system operation. It is required to process monthly reports sent by employers on members' salaries paid, pension contributions made, and services provided in order to maintain an up-to-date record of all the active members. It is required to process benefit payment checks and make direct deposits of monthly benefit payments to retirees. It is also required to support the production of system reports, system studies, and system control.
2. Accounting and financial services: The staff performs the accounting and financial services of the pension system, such as recording and depositing of contributions made by members and employers to the pension system, daily accounting of the invested assets of the system, and preparation of the annual financial statement of the pension system.
3. Legal services: Legal services provided by the pension system staff are also essential in running the pension system within the legal framework. Due to the importance of fiduciary duties in pension management, the legal staff will have a responsibility to provide advice to the board of trustees on the fiduciary duties. The legal staff also interprets legislation related to retirement that affects the pension systems' members and plan sponsors.
4. Actuarial service: Most large pension systems hire outside actuaries to provide this critical service. A few state pension systems, such as New York and

Washington state pension systems, have their own in-house actuaries to conduct actuarial valuations.

4.3.2.3 Investment Management

The staff is involved in various aspects of investment management, from advising the board on investment policy to selecting investment managers and managing investment. This topic is covered in more detail in the next chapter.

4.3.3 Risk Management and Control

Due to the vast assets under management, the long-term nature of pension benefits and investment, and the numerous parties involved, pension plan administration is fraught with risks, more so than other aspects of public financial management. There needs to be a risk management and control mechanism in place to ensure that all persons or entities with operational and oversight responsibilities act in accordance with the objectives set out in the pension entity's bylaws, statutes, and policies. This risk management and control mechanism is maintained by both internal staff and external professionals. Externally, it means an independent auditing of the system's financial statements by certified public accountants. Internally, while the governing board, the executive director, and the staff are all responsible for internal control, a linchpin in the implementation of a more comprehensive internal risk management system is internal auditing by an internal auditor. Internal auditors serve many functions, from being a watchdog over the management of the pension plan to teaching board trustees and staff about pension management.

For example, in 1995, Wisconsin Act 274 created an internal audit function within the board. Directed by the internal auditor, the internal audit unit may review any activity of the board and has access to the records of the board and any external party under contract with the board (State of Wisconsin, 2006). The auditor plans and conducts audits, risk assessments, research projects, and management reviews under the direction of the board; assists with external audits and reviews of the board; and monitors the board's contractual agreements with financial institutions, investment advisors, and any other party providing investment services to the board. By directly reporting to upper-level management and fund trustees, internal auditors advise decision makers about potential problems and the ways to correct them. Therefore, compared to an external auditor, an internal auditor has a more intimate knowledge of and also exerts greater impact on the operation of the whole pension system.

Internal auditors belong to an organization called the Association of Public Pension Fund Auditors (APPFA). APPFA was formed in Chicago in 1991 by four internal auditors from pension systems in Colorado, Illinois, New York, and Wisconsin. Since then, APPFA has grown to 72 members from the United States and

Canada, including most of the large state and local pension systems in the United States.*

In 2000 and 2003, APPFA published two documents titles *Public Pension Systems: Statements of Key Investment Risks and Common Practices to Address Those Risks* and *Operational Risks of Defined Benefit and Related Plans and Controls to Mitigate Those Risks*.† These two documents systematically examine all the major risks facing public pension systems and the mechanisms in managing such risks.

4.3.4 Financial Reporting

Even though public pension systems are not subject to federal regulation on financial reporting, they are nonetheless subject to substantial reporting requirements by state statutes and the Governmental Accounting Standards Board (GASB). First and foremost, the pension system is required to publish an annual comprehensive financial report (CAFR), prepared based on the standards set in GASB 25. The CAFR is divided into four sections: financial, investment, actuarial, and statistical:

- The Financial section begins with a management discussion and analysis (MD&A), which explains the main operational results of the pension plan in the past year and alerts the readers to any major events and changes that will have an impact on the plan in the future. The bulk of the Financial section consists of the two statements (Net Assets and Changes in Net Assets), notes to the statements, two schedules (Funding Progress and Contributions from Employer), and notes to the schedules, as discussed more fully in Chapter 3.
- The Investment section contains information on the plan's asset allocation, current and historical investment returns, external investment managers, and their fees.
- The Actuarial and Statistical sections contain more information on the actuarial valuation of the pension plan and historical trends, such as benefits paid.

The CAFR also has to be authenticated by an outside auditor who issues a statement of opinion as to whether or not the financial statements and schedules are presented fairly and in accordance with generally accepted accounting principles. All pension systems' CAFRs are available to any citizen upon request and most of them can be found on pension systems' Web sites as well.

---

* Association of Public Pension Fund Auditors. *The Insiders Who Audit Public Pension Funds.* http://www.appfa.org (Accessed 5/16/2007.)

† Both reports are available at APPFA's Web site at http://www.appfa.org/

The public pension plan sponsor also has to disclose information on the pension plan according to the standards set in GASB 27, as discussed in the previous chapter. In addition to the annual financial report, most public pension systems are also required to submit reports to a legislative body and, in some cases, government agencies created by plan sponsors to oversee the pension systems.

4.3.5 Funding Policy

While public pension plans show little variation in the various aspects of plan administration discussed so far and conform substantially to the standards set by ERISA, funding policy is one aspect of pension plan administration that displays more substantial variation among public pension plans and deviation from ERISA standard for some plans. Funding policy refers to the method used by the pension plan sponsor to determine the periodic contribution it has to make to the pension plan so as to accumulate sufficient assets for paying future pension benefits. ERISA requires that pension plans be funded on an actuarial basis, meaning the periodic contribution to the pension plan should include normal cost plus an amount to amortize the unfunded pension liability, with the maximum amortization period set to 30 years.

An examination by the author of the funding policy of all major state-level pension plans found that there are three different types of funding policy in the public sector.* The first type of funding policy is the same as that required by ERISA. State pension plan sponsors in 34 states have adopted the funding policy that requires pension contributions be determined actuarially and the pension plan sponsor should pay fully the amount determined actuarially. While the vast majority of the plan sponsors with this funding policy pay the full amount, the funding policy by itself does not necessarily guarantee that the full actuarial amount will always be paid. For example, even though the Kentucky state government is required to contribute at an actuarially determined rate, it significantly underfunded its pension contribution to the Kentucky Employees Retirement System from fiscal year 2004 through 2007 (Kentucky Retirement System, 2007).

The second kind of funding policy is similar to the first kind, but with some adjustment or flexibility built into it. Such a funding policy is used by plan sponsors in four states:

1. **Alaska**: The employer contribution rate is determined actuarially. However, state regulation 2AAC 35.900 prohibits the rate from going up or down by more than five percentage points from the rate adopted in the prior year.

---

* This examination is conducted through a review of each pension plan's comprehensive annual financial report, which is required by Governmental Accounting Standards Board (GASB) to disclose its funding policy.

2. **Kansas**: The employer contribution rate is determined actuarially. However, there is a statutory cap on the increase in the contribution rate from the prior year, set to 0.6 percent in fiscal year 2008 and beyond.
3. **Massachusetts**: Chapter 32 of the General Laws directs the secretary of administration and finance to prepare a funding schedule to meet actuarially determined requirements and to update this funding schedule every three years on the basis of new actuarial valuation reports prepared under the secretary's direction. Any such schedule is subject to legislative approval. If a schedule is not so approved, payments are to be made in accordance with the most recently approved schedule.
4. **New Jersey**: The employer contribution rate is determined actuarially. However, the rate can be amended by state legislation.

The third kind of funding policy, found in 10 states, is loosely linked to the actuarially based funding policy. In these states, the pension contribution rate is set by the state government through legislation. The ability of a plan sponsor to set a contribution rate through legislation is one of the most critical differences between public and private pension plans. As a public pension plan sponsor is also a legislative body, and since there is no federal regulation of public pension plans, it is thus unavoidable that some plan sponsors will use legislative power to set the contribution rate.

Statutory contribution rate by itself does not necessarily mean that it is substantially different from the actuarially determined rate. Those states that set statutory contribution rates can be divided into two groups, depending on the circumstances under which statutory rates are set.

In the first group, the statutory rate is initially set at a level that is linked to the actuarial rate, meaning that if the statutory rate is met every year and all the actuarial assumptions are met, the statutory rate is sufficient to fully fund the pension benefits. The purpose of setting the statutory rate is to have a more stable contribution rate over time as the actuarial rate can change depending on the funding ratio of the pension plan. This means that the statutory rate can be higher or lower than the actuarial rate from time to time. Examples of states using such funding policy are Wyoming, Texas, Iowa, Minnesota, Nevada, Colorado, and Connecticut. The statutory rate does not change unless it deviates substantially from the actuarial rate, usually as a result of a significant drop in pension funding ratio and the present statutory rate leads to an amortization period much longer than the maximum 30 years. For example:

1. In Iowa, the statutory contribution rate remained unchanged from 1979 to 2007. The pension plan was near full funding in 2000. Due to a drop in funding ratio to 88 percent in 2006, the Iowa State Legislature passed legislation to increase the contribution rate. The increase of two percentage points from 9.45 percent to 11.45 percent will be phased in over four years beginning July 1, 2007 (Iowa Public Employee Retirement System, 2006).

2. In Colorado, members and employers are required to contribute to Public Employees' Retirement Association (PERA) at a rate set by statute. On December 31, 2005, the state division of PERA had a funded ratio of 71.5 percent. In the 2004 legislative session, the Legislature passed Senate Bill 04-257, which established Amortization Equalization Disbursement (SAED). The Bill requires PERA employers to pay an additional 0.5 percent of total salaries paid beginning January 1, 2006, increasing by 0.5 percent in 2007 and by 0.4 percent of salary each subsequent year, reaching a maximum of three percent in 2012 and thereafter. This payment will be used to pay for unfunded liability and will be terminated once the unfunded liability is eliminated (Colorado PERA, 2007)

In the second group, which includes Illinois, Oklahoma, and West Virginia, the state government sets the statutory rate to correct severe underfunding. In these states, the state government has been contributing substantially below the actuarial rate for a long period of time so that the pension plan is severely underfunded. Facing very low funding ratio and sometimes court ordered to correct the funding situation, the state government has been forced to increase contributions to bring the pension plan to full or near full funding status over a period of time. To avoid the shock to the government budget, the state sets the statutory rate and gradually increases it to bring it close to or above the actuarial rate. All of these three states are in the midst of a multidecade funding schedule to bring the funding ratio to 80 or 90 percent.

Of the 49 states that have state-level defined benefit pension plans,* Indiana is the only state that funds one of its two state-level pension plans partially on a pay-as-you-go basis. Indiana State Teachers' Retirement Fund (TRF) is funded on a pay-as-you-go basis for employees hired prior to July 1, 1995. State appropriations are made for the amount of estimated pension benefit payout for each fiscal year. If the actual pension benefit payout for the fiscal year exceeds the amount appropriated, the difference is paid from the Pension Stabilization Fund. For employees hired on or after July 1, 1995, the individual employer will make annual contributions that are actuarially determined. Due to the partial pay-as-you-go funding method, the funding ratio of TRF improved very slowly. Over a 10-year period from 1996 to 2005, the funding ratio increased from 31.6 to 44.8 percent.

4.4 Public Pension Plan Oversight

In the private sector, the Employee Benefits Security Administration of the Department of Labor is responsible for the administration and enforcement of Title I of

* As explained in Chapter 7, Nebraska is the only state that does not have a state-level defined benefit plan.

ERISA. Because the Congress exempts state and local pension plans from ERISA, no federal government agency has oversight and regulatory authority over public pension plans other than the Internal Revenue Service, which determines the tax-exempt status of public pension plans. Despite this lack of oversight from the federal government, state governments, which are the plan sponsors of all the large public pension plans, have vested various institutions with oversight authority over public pension plans, although some have more authority than others. Such oversight authority is generally vested with two types of government entities: state legislative committee or independent pension commission created by the state legislature.

4.4.1 Legislative Committee

Since a state-level public pension system is created through state legislation, the state legislative body has the ultimate oversight authority over the pension system. Legislative committees that oversee pension systems can be divided into two general groups. In the first group, many states do not have a specific committee dealing with public pension issues. Thus, the oversight authority generally falls under committees that deal with overall public financing issues, such as finance or ways and means committees. In the second group, still a significant number of states establish legislative committees dedicated to pension financing issues. These committees vary in the scope of their responsibilities. Some are responsible for pension policies and state laws governing pension systems, and others have more direct authority over the administration of pension system. Some examples of state legislative pension committees include: Indiana Pension Management Oversight Commission (PMOC), Louisiana Joint Legislative Retirement Committee, Minnesota Legislative Commission on Pensions and Retirement (LCPR), North Carolina Standing Committees on Pensions and Retirement, and Wisconsin Joint Survey Committee on Retirement Systems.* A brief description of Indiana PMOC and Minnesota LCPR shows the general structure and responsibilities of such legislative pension committees:

- Indiana PMOC was created in 1985. It consists of four members from the Senate and the House of Representative each, and four lay members. The statutory duties of the commission include: (1) studying the investment and management practices of the boards of the public retirement funds; (2) determining what constitutes adequate wage replacement levels at retirement (including benefits from public retirement funds and social security) for public employees; (3) studying the impact of federal law and proposals concerning pensions, annuities, and retirement benefits; (4) studying the retirement

* For all state legislative committees on pension-related matters, please see Moore (2005).

funds established in IC 36-8; and (5) studying methods and levels of funding for public retirement funds (Indiana Legislative Services Agency, 2006).

- Minnesota LCPR makes recommendations to the legislature including financing of the various pension funds and financing of accrued liabilities. It oversees over 700 state and local plans. The commission has fourteen members, five members from the House and Senate each and four staff members. The larger plans are required to submit an actuarial valuation to the LCPR. The LCPR sets the guidelines for actuarial assumptions used in the valuations. Plans' investments are monitored by the state auditor, in conjunction with the LCPR.

4.4.2 Independent Pension Commission

An independent pension commission is different from legislative pension committees in two important ways. First, the membership is different. The majority of members on a legislative committee are legislators, whereas the majority of members on an independent pension commission do not come from the legislative body. The membership of an independent commission typically represents a broader scope of interests and usually one or more members are required by the enabling statute to have expertise in pension financing. The size of independent commission, including both members and staff, is also much larger than that of legislative committee. The larger size of independent pension commission also brings it more responsibilities. Second, the source of financing for the activities of an independent commission can come from the pension systems themselves without legislative appropriation.

There are many reasons why an independent pension commission can be advantageous to a legislative committee. The main reason is the short-term and political nature of the budget cycle and the long-term nature of pension funding. Politicians tend to think of public financing issues in terms of the immediate cost over the next budget cycle, lasting one to two years. As discussed in more detail in Chapter 6, pension policy can be easily influenced by short-term budgetary concerns. Pension financing, however, is very long-term in nature, with the cost distant in the future. This makes long-term planning and analysis far more important for pension than for most other government programs and calls for more consistent policy guidance in the long run. Another reason is that pension is a very complex subject that requires a high degree of expertise and knowledge. A permanent independent pension commission, thus, can give a state legislature a more consistent independent source of information and policy guidance regarding public pension financing issues. There are five state-level independent pension commissions with a broad scope of responsibilities in Massachusetts, Ohio, Oklahoma, Pennsylvania, and Texas. Following is a brief discussion of three of these permanent pension commissions.

1. Massachusetts Public Employee Retirement Administration Commission (PERAC): PERAC was created in 1996 to oversee, guide, monitor, and regulate 106 Massachusetts public pension systems. It consists of seven members, with three appointed by the governor, three by the state auditor, and one chosen by the first six members. Of the three persons appointed by the governor, one is the governor or his designee, one is a representative of a public safety union, and one is qualified by having training and experience in the investment of funds for at least ten years. Of the three persons appointed by the state auditor, one is the state auditor or his designee, one is the president of the Massachusetts AFL–CIO or his designee, and one is a representative of the Massachusetts Municipal Association. The commission has approximately fifty staff members in nine units, including actuarial, legal, audit, investment, disability, and fraud. The commission monitors disability claims, investigates fraud, performs actuarial valuations and experience studies, and conducts audit reports for the state's pension plans. The members serve without compensation and the budget for the commission is funded from the investment income account of the state retirement systems.
2. Ohio Retirement Study Council: The Ohio Retirement Study Council (ORSC) was created in 1968.* The Council is composed of fourteen members: three members of the House; three members of the Senate; three members appointed by the governor, one representing the state, one representing local governments, and one representing public education institutions; and the five executive directors of the state retirement systems, who are nonvoting members. Council members serve without compensation and the budget for the council is paid out of the investment earnings made on the assets of the five state retirement systems. The council receives no legislative appropriations and performs the following statutory duties:
    1. Makes a review of all laws governing the public retirement systems and makes recommendations to the legislature on any changes with respect to benefits, sound financing of benefit costs, and prudent investment of funds.
    2. Reports to the governor and legislature on its evaluation and recommendations with respect to the operations of the public retirement systems and their funds.
    3. Studies all proposed changes to the public retirement laws and reports to the legislature on their costs, actuarial implications, and desirability as a matter of sound public policy.
    4. Reviews semiannually the policies and objectives of the systems' investment programs.

* Ohio Retirement Study Council. *About ORSC.* http://www.orsc.org/aboutorsc.cfm.

5. Prepares, at least once every ten years, an independent actuarial review of the annual actuarial valuations and quinquennial actuarial investigations prepared by each system.

3. Texas Pension Review Board (PRB): The PRB was created in 1979 as an independent state agency to oversee and review state and local government retirement systems in Texas. The board is composed of nine members, appointed by the governor, the lieutenant governor and the speaker of the House. The board employs an executive director to be the executive head of the board and perform its administrative duties. The board is financed by a special fund created in the state treasury, with the funds coming from both legislative budget appropriation and contribution from the public pension systems in Texas. The board's responsibilities include:

    1. Conducting a continuing review of all public retirement systems within the state, compiling and comparing information about benefit structures, financing, and administration of systems
    2. Conducting intensive studies of existing or potential problems that weaken the actuarial soundness of public retirement systems
    3. Recommending policies, practices, and legislation to public retirement systems and their sponsoring governments
    4. Examining all legislation for potential effect on Texas' public retirement systems, overseeing the actuarial analysis process, and providing actuarial review when required by law

4.4.3 Other Oversight Mechanisms

For those states that do not have an independent pension commission or standing legislative committee on pension issues, they also form temporary pension commissions from time to time to study pension-related issues. They are temporary because they exist for only a short period of time, usually about one year. Compared to a permanent commission, a temporary commission is usually charged with limited authorities, with the main purpose to review current policies and practices and to make recommendations on pension reforms. A temporary pension commission is formed usually at a time when the public pension system is facing a severe long-term funding shortage and major reform is needed to put the pension system on a more sustainable path. For example, the Michigan Commission on Public Pension and Retiree Health Benefits was created in 1999 to (1) review those state laws that govern or affect the funding, management, oversight, and fiscal integrity of public pension and retirement systems; (2) review the adequacy of funding for public pension and retirement systems and the extent of unfunded accrued liabilities; and (3) consider, recommend, and report such modifications in state laws governing or affecting public pension and retirement systems. The commission consisted of nine members appointed by the Governor and had to complete its work not later than one year after the commission was appointed.

In the early 2000s, after the severe stock market downturn that led to a decrease in the funding ratio of pension plans, many states, such as Illinois, New Jersey, and California, formed pension commissions to study pension financing issues. However, the recommendations by the commissions are usually not binding on the state legislature. Despite the lack of enforcement authority, temporary pension commissions still achieve the purpose of alerting the elected officials and the public to the important systemic and policy issues facing the pension plans.

Another state oversight mechanism is auditing. In some states, state legislatures and state agencies have direct auditing authority over public pension systems. For example, the Virginia Retirement System (VRS) Oversight Act (Section 30-78 et seq. of the Code of Virginia) directs the Virginia Joint Legislative Audit and Review Commission (JLARC) to be responsible for continuing oversight of the Virginia Retirement System. JLARC is required to publish periodic status reports and semiannual reports, which summarize the performance of VRS investments. In Wisconsin, the Legislative Audit Bureau conducts a financial audit of Wisconsin Retirement System, including an assessment of the fair presentation of the financial statements. The audit also evaluates the board's internal controls and compliance with applicable statutes, policies, and guidelines. The Legislative Audit Bureau conducts a biennial performance evaluation that includes an audit of the board's policies and management practices.

In Minnesota, the Office of State Auditor monitors investment, financial, and actuarial reporting for over 700 public pension funds. Each year, public pension plans with a market value of less than $10 million in assets are required to report to the State Auditor's Office.

References

California State Teachers' Retirement System. 11/3/2006. *California's Teachers' Retirement Board Sets Higher Ethics Bar.* http://www.calstrs.com/newsroom/2006/news110306.aspx (Accessed July 8, 2007.)

Colorado Public Employees' Retirement Association. 2007. *Comprehensive Annual Financial Report for the Fiscal Year Ended December 31, 2006.* http://www.copera.org/pdf/5/5-20-06.pdf (Accessed August 8, 2007.)

Indiana Legislative Services Agency. 2006. *Annual Report of the Pension Management Oversight Commission.* http://www.in.gov/legislative/interim/committee/reports/PMOC9B1.pdf (Accessed July 20, 2006.)

Iowa Public Employee Retirement System. 2006. *Comprehensive Annual Financial Report for the Year Ended June 30, 2006.* http://www.ipers.org/publications/misc/pdf/financial/cafr/cafr2006.pdf (Accessed June 3, 2007.)

Kentucky Retirement Systems. 2007. *Comprehensive Annual Financial Report for Fiscal Year Ended June 30, 2007.* http://www.kyret.com/cafr/cafr2007.pdf (Accessed January 24, 2008.)

Moore, Cynthia. 2005. *Protecting Retirees' Money*, 5th ed. National Council on Teacher Retirement, Sacramento, CA.

National Education Association, *Collective Bargaining and Member Advocacy*. 2006. Characteristics of Large Public Education Pension Plans. http://www.nea.org/takenote/images/char2006.pdf (Accessed June 2, 2007.)

Ohio Public Employees Retirement System. 2005. *Board Governance Policy Manual.* http://www.nasra.org/resources/OH%20PERS%20Governance%20Manual.June%202005.pdf (Accessed June 3, 2007.)

Ohio Ethics Commission. 5/12/2006. Press release: *Former STRS Board Member Sentenced for Ethics Violations.* http://www.ethics.ohio.gov/PressReleases/05122006.html (Accessed July 23, 2007.)

———9/19/2006. Press release: *Four STRS Board Members Convicted of Ethics Charges.* http://www.ethics.ohio.gov/PressReleases/09192006.html (Accessed July 23, 2007.)

State of Wisconsin, Legislature Reference Bureau. 2006. *Oversight of Wisconsin Retirement System.* http://www.legis.state.wi.us/lrb/pubs/ib/06IB2.pdf (Accessed August 9, 2006.)

U.S. General Accounting Office. 1979. *Funding of State and Local Government Pension Plans: A National Problem.* (H RD-79-66), Washington, D.C.

U.S. House of Representatives Committee on Education and Labor. 1978. *Pension Task Force Report on Public Employee Retirement Systems*, Washington, D.C.

Chapter 5

Investment Management

Investment is by far the most important part of pension plan management. As discussed in Chapter 1, there are three major sources of income to pay for promised retirement benefits, employer and employee pension contributions, and investment income. Table 1.5 in Chapter 1 also indicates that, over time, investment income far outweighs pension contributions as a funding source for pension benefits. As the ultimate responsibility of any pension system is to ensure that there will be sufficient assets in place to pay pension benefits, investment management, therefore, is the top priority of the pension system and the governing board that is set up to manage the pension system.

5.1 Overall Objective of Investment Management

The goal of pension plan investment is to achieve a long-term rate of return that can generate sufficient investment income. This rate of return should be at least equal to the assumed rate of return used to discount future pension benefits. As the average assumed rate of return is about 8 percent, higher than the return on risk-free Treasury securities, achieving such return involves taking on investment risk. Investment risk refers to the uncertainty of achieving the desired return in the future. The fundamental principle of investment is the tradeoff between return and risk. The less the uncertainty about the future return, the lower the future return will be, and vice versa. This uncertainty, or investment risk, is traditionally measured by the standard deviation of the periodic return around the historical average (or mean) return.* The greater the standard deviation, the greater the uncertainty of not achieving the average return in any given investment period. According to Ibbotson Associates (2006), over an 80-year period between 1926 and 2005, the average returns for small company stocks, large company stocks, long-term government bonds, and U.S. Treasury bills were 12.6 percent, 10.4 percent, 5.5 percent, and 2.8 percent, respectively, and the standard deviations of returns for these four types of securities were 32.9 percent, 20.2 percent, 9.2 percent, and 3.1 percent, respectively.

It is obvious that over a long period of time, there is a strong positive correlation between return and risk. The only way for a pension plan to eliminate any investment risk is to invest in securities with relatively low return. Such an investment strategy, however, creates another type of risk, namely the risk of not generating enough investment income over time and causing a significant increase in employer and employee contributions. If the pension plan pursues a high return/risk investment strategy, it then also runs the risk of creating substantial ups and downs in asset value from year to year and unpredictable pension contributions, highly undesirable from a government budgeting perspective. Therefore, the overall objective in pension asset investment is to achieve a desired rate of return while minimizing investment risk.

5.2 Portfolio Diversification

In this section, we first look at the modern portfolio diversification theory and then examine the various asset classes that provide such diversification.

5.2.1 The Theory

The solution to achieving the objective mentioned above is found in the modern portfolio diversification theory. To understand the portfolio theory, we need to first understand the source of investment risk. The risk of an investment can be divided into two components: systemic risk and nonsystemic risk. Systemic risk refers to the investment risk due to the overall investment environment. This can be caused by the overall economic conditions, such as the economy performing very differently from expected or the inflation rate being different from expected. These are factors that are common to investment in every financial security. Investors cannot avoid taking on that risk and, thus, should be rewarded with a risk premium for doing so.

* For example, if an investment instrument has an average return of 10 percent and a standard deviation of 20 percent, then there is a 67 percent probability that the future return will be between –10 percent and 30 percent.

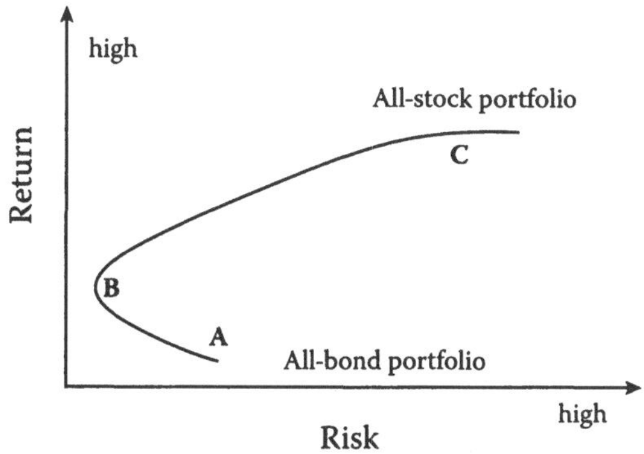

Figure 5.1 Risk/return profiles of a hypothetical investment portfolio.

Nonsystematic risk, however, is the risk that is unique to a particular investment security like an individual stock. This unique risk can be due to the product or service produced by a company, or it can be due to the company's management skill. By investing in only one stock, the investor is bearing both the systemic risk and nonsystemic risk. This nonsystemic risk, however, can be reduced by adding more and more stocks to the investment portfolio. Since each stock has its own unique risk, the unique risks of all the stocks can be offset from each other as long as the risks are not highly related to each other. If there are sufficient stocks in the portfolio, the nonsystemic risk can be eliminated, leaving investors with only the systemic risk.

The principles of portfolio diversification theory were first introduced in 1952 by Harry Markowitz who theorized that the inherent risk associated with individual stocks could be offset by including assets that have significantly different return and risk profiles. Two assets have different risk profiles when the price for them does not rise and fall roughly in tandem. In this case, the unique risk of the two assets is offset by each other, resulting in a reduction in the risk of the portfolio of these two assets.*

For example, suppose an investment portfolio consists of only two assets, bonds and stocks. Bonds have a lower return, but also a lower risk compared to stocks. Historically, the returns of bonds and stocks do not always go hand in hand and many times they go in opposite directions. Figure 5.1 illustrates the risk and return profile of this portfolio whose composition goes from all bonds to all stocks. Point A represents a portfolio that consists of only bonds. When the composition of the portfolio shifts gradually toward stocks, the overall risk of the portfolio is reduced

* The measurement of this relation between the two securities is called correlation coefficient. It is between –1 and 1, when –1 means that assets are perfectly uncorrelated and 1 means perfectly correlated. A lower correlation reduces the risk, or standard deviation of the portfolio, thus providing the benefit of diversification. Therefore, a diversified portfolio should consist of asset classes that have low correlation among them.

and the return is increased at the same time. The portfolio at point B has the lowest risk. The portfolio at this point should be preferred to any point below it on the curve because it represents lower risk, but also higher return, as the result of portfolio diversification. This shows the benefit of having two assets whose price changes are not correlated with each other: lower risk and higher return at the same time. When the portfolio shifts gradually from point B to point C, the return is increased along with the increase in risk.

When more and more assets with different risk and return profiles are added to the portfolio, the unique risks of the individual assets continue to be diversified away and the risk of the overall portfolio continues to be reduced while the return of the portfolio is increased. With a different combination of all available assets, an efficient frontier can be created. Each point on the frontier represents an optimal portfolio. An optimal portfolio means that given a certain return objective, this portfolio will produce the lowest level of risk, or given a certain level of risk, this portfolio will produce the highest rate of return. Any portfolio below the frontier is inefficient because another portfolio can be found to produce the same return with a lower level of risk. Which optimal portfolio on the efficient frontier an investor will choose depends on the investor's return objective and risk tolerance.

5.2.2 Asset Classes

To arrive at such an optimal portfolio, it is necessary to know what investable securities are available and how the return of one security correlates with all other securities. Since there are tens of thousands of investable securities (the U.S. stock market alone has over 7000 stocks), it is a very demanding task to calculate the correlation of each pair of securities in order to arrive at an optimal portfolio, even with the help of modern computing technology. To make the task more manageable, individual securities are grouped into asset classes, such as stocks and bonds. An asset class includes securities that have a similar risk and return profile. As there still exists substantial variation in the risk and return profile of individual assets in one asset class, each asset class is usually further divided into subclasses to arrive at a more homogenous risk and return profile within each subclass. Instead of collecting data on the risk and return profile of each individual security and the correlation of each pair, we now only need to know the risk and return profile of each asset class and its subclasses as well as the correlation between different asset classes in order to arrive at an optimal portfolio. Therefore, knowing all the major asset classes is the first step in understanding pension asset investment.

Asset classes can be divided into two broad categories: traditional investments and alternative investments.

5.2.2.1 Traditional Investments

There are four main asset classes in the traditional investments category: equity, fixed-income securities, real estate, and cash or cash equivalents. In this category, the various investment securities are traded on public exchanges, making them easy to buy and sell. Both the values of securities and changes in the value are easily observable due to frequent trading of these securities.

5.2.2.1.1 Equity

Equity represents ownership of a corporation. The return on equity comes from dividends and the increase in the price of equity. Equity has the highest risk among the traditional asset classes. The return is not known at the time of purchase. At its worst, the value of the stock can go down to zero if the company is liquidated. In the best case scenario, there is theoretically no limit as to how high the price can go if the corporation becomes a huge success. While risky compared to other traditional asset classes, equity also offers the best long-term investment return. Not all stocks, however, are created equal. Stocks can be further divided into many subclasses based on geographic location, market capitalization, and style. Further classification is necessary to arrive at a more precise risk/return profile of various segments of the equity market.

In terms of geographic location, stocks are divided into U.S. domestic stocks and international stocks. International equity provides a source of diversification for the investment portfolio because historically the return on international equity is comparable to domestic equity, but these two returns are not highly correlated. The return on international equity is also affected by currency exchange rates. Since the exchange rate fluctuates from year to year, the actual return in the U.S. dollar from international equity thus can be quite different from that of U.S. domestic equity. International stocks can be further divided into those from developed countries, such as Japan and western European countries, and those from less developed countries, such as India, China, Russia, and countries in Latin America. The stock market in less developed countries is also called the emerging market. The distinction between international developed stock market and emerging market is that the latter is more risky and the return is also potentially higher due to the faster economic growth rate.

Stocks can also be divided into three subclasses depending on market capitalization: large cap, mid cap and small cap.* Large caps are stocks of large and well-established companies, whereas small caps are stocks of much smaller companies. The distinction between large and small caps is that the price of small stocks is more volatile than that of large stocks, but over time the return on small stocks is

---

* Market capitalization is the market valuation of the company, determined by the current price of its stock and the number of outstanding shares.

also higher than that on large stocks. Stocks can also be divided into two subclasses based on style, growth, and value, determined mainly by their price/earning (P/E) ratios.* The higher the P/E ratio, the more the investor is willing to pay for every dollar of profit in the future. Growth stocks have higher P/E ratios, indicating investors expect higher growth potential for the companies and are willing to pay a premium for such potential. Value stocks tend to have below average P/E ratios. The distinction between growth and value stocks also has to do with risk and return profile. Because of the higher P/E ratio, the price of growth stocks is more volatile than that of value stocks, but at the same time the return on growth stocks is also expected to be higher than that on value stocks.

5.2.2.1.2 Fixed-Income Securities

Fixed-income securities, more commonly known as bonds, pay a fixed interest rate and because they have senior claim on a company's assets if they are liquidated in the case of bankruptcy, they are less risky than stocks and, thus, have on average a lower return than stocks.

Based on default risk, bonds can also be divided into three subclasses: high, medium, and low credit quality, based on the bonds' credit rating. Bonds with a high credit rating have low default risk and also a lower return. A special type of low-quality bond is called high-yield bond, or junk bond. High-yield bonds are those with credit ratings that are below investment grade.† They have a much higher risk of default and, thus, carry a return that is also much higher than investment-grade bonds.

In addition to the default risk, another source of risk when investing in bonds is the interest rate risk. The price of bond and interest rates moves in opposite directions. When interest rate goes up, the price of a bond goes down. Based on the interest rate risk, bonds can be divided into three subclasses depending on their maturity: long-, intermediate-, and short-term. For the same change in interest rate, bonds with longer maturity witness a greater percentage change in price and, thus greater, volatility. In return, bonds with longer maturity also have higher return. The maturity of a bond or a portfolio of bonds is measured by duration, which is the weighted average life of a bond or a portfolio of bonds. The longer the duration, the more risky the bond or the portfolio of bonds is.

Like stocks, bonds can also be divided into domestic bonds and international bonds, and international bonds can be subdivided into bonds issued by developed and developing countries. Developing country bonds have higher credit risk, but also carry higher return compared to those issued by developed countries.

---

* Price/earning ratio is calculated by dividing the price of the stock by the earnings per share of the stock.

† Crediting ratings that are of investment grade are AAA, AA, A, and BBB. Anything rating below BBB is of noninvestment grade.

Bonds provide an important source of diversification in an investment portfolio due to the low correlation with stocks. The return on bonds increases when the interest rate decreases. Interest rate decrease usually happens at a time when the economy slows, which is more likely to have a negative effect on the return of stocks. Therefore, the low return on stocks can be offset to some extent by the high return on bonds, reducing the volatility of the overall portfolio.

5.2.2.1.3 Real Property

Return on real property investment is generated from both income in the form of rent and appreciation in the value of real property when it is sold. There are two principal ways of investing in real estate. One way is for the pension plan to directly purchase income-producing properties, such as office buildings, shopping malls, apartment buildings, and warehouses. There are three main drawbacks of this method. First, the number of properties that can be bought in such a way can be limited and the real property portfolio is not well diversified. Any loss of value in one property can have a severe impact on the return. Second, selling and buying individual properties is not as easy as other traditional assets, thus creating a liquidity risk for the portfolio. Third, there is substantial cost involved in managing properties for both time and manpower.

A second more popular way is to invest in a real estate investment trust (REIT). A REIT is a company that owns and operates income-producing real estate properties. To qualify as a REIT, a company must distribute at least 90 percent of its taxable income to its shareholders annually. The shares of REITs are traded on the stock market, just like stocks of other companies. Compared to directly owning properties, REITs provide greater diversification by pooling together investors' funds and investing in a portfolio of properties. REITs add significant diversification to an investment portfolio. The return on REITs is comparable to stocks with a fairly low correlation with the overall stock and bond markets. For calendar years 2000, 2001, and 2002, for example, the U.S. stock market suffered heavy losses, whereas the price of REITs increased substantially over this period. Such negative correlation reduces the volatility of the portfolio and boosts the long-term return of the portfolio.

5.2.2.1.4 Cash and Cash Equivalents

Cash or cash equivalents refer to short-term investments (with maturities less than one year) with a fixed interest rate. Due to the very short-term nature of these securities, they carry very little or no risk and, thus, their returns are also much lower than those on bonds and stocks. Securities included in this asset class are Treasury bills, money market funds, and certificates of deposit, among others. As the return on this asset class is fairly low, it usually accounts for a relatively small percentage of a pension portfolio that is invested for the long run. Investment in cash equivalent asset class is not so much for diversification purposes as it is for liquidity purposes. Having cash equivalent assets in the portfolio allows the pension plan to generate cash quickly without loss of principal and earn some investment income at the same time.

5.2.2.2 Alternative Investments

There are three main asset classes in the alternative investment category: private equity, hedge funds, and real property investment through a private investment group. Unlike the traditional investments, investment instruments in this category are not traded on any public exchange. They are usually purchased through a limited partnership arrangement. Investment in commodities, such as timber and gold, while rare among public pension systems, is another type of alternative investment. The benefits of alternative investments include potentially higher returns and reduced volatility, resulting from low correlation with traditional investments. Alternative investment is different from traditional investment in three key aspects. First, alternative investment is less transparent than traditional investment in terms of valuation due to the lack of trading. Second, alternative investment is less liquid as the investment is usually locked up over a period of many years and early withdrawal can result in a substantial loss of principal. Third, the investment cost is also much higher for alternative investments, due to the high level of expertise needed in identifying and managing such investments.

5.2.2.2.1 Private Equity

Private equity refers to equity investment in a company that is not listed on a public stock exchange. For a public pension plan, such investment is done through investment in a private equity fund that pools the resources for such investment. Generally, private equity funds are organized as limited partnerships, which are controlled by private equity firms.

Private equity is a broad term for several very different types of investment, such as venture capital and buyout. A venture capital fund is a pooled investment vehicle that provides financial capital for new and growing businesses. Buyouts provide leveraged capital to take a public firm private in order to restructure the firm. Whether it is venture capital or a buyout, the return on private equity investments is typically realized in the form of capital gains through the sale of the company, or by eventually letting the company go public on the stock market.

While private equity carries a similar risk to that of public stocks, such as losing all the money if the private equity funds invest in companies that fail, it is regarded as a distinct asset class. Private equity has historically outperformed traditional asset classes and is only moderately correlated to the stock and bond investments

(Center for International Securities and Derivative Markets, 2006), thus providing further opportunity to diversify a portfolio beyond traditional securities.

5.2.2.2.2 Hedge Fund

Even though hedge funds have been in existence since the middle of the twentieth century, they did not become a popular alternative investment asset class until the early 2000s. Hedge fund is a term that previously described a type of private investment fund that employed hedging and arbitrage techniques in the corporate equity market. The funds used leverage and short selling to "hedge" the portfolio's exposure to movements of the corporate equity markets.* Over time, hedge funds have gradually evolved to diversify their investment portfolios to include other financial instruments, such as fixed income securities, currencies, exchange-traded futures, derivatives, futures contracts, and commodity options. They also engage in more investment strategies in addition to hedging and arbitrage. Today, the term "hedge fund" refers not so much to investment techniques as to funds that are private and unregistered (Securities and Exchange Commission, 2003).

The various investment strategies, such as hedging and arbitrage, are generally not available to managers of traditional mutual funds in stocks and bonds. Because of the different investment strategies involved, the return on hedge funds has low correlation with returns on traditional stocks and bonds. Therefore, adding a hedge fund to the investment portfolio provides further diversification.

5.2.2.2.3 Real Property

Investment in real property outside REITs qualifies as an alternative investment for two reasons. First, it is managed by a private group of professionals who buy, manage, and sell commercial properties. An example of this is the JP Morgan Asset Management Group, which manages several real estate funds that invest in the United States, Europe, and Asia. Unlike REITs, such investment funds are not traded publicly and, thus, there is usually restriction on how often the investment can be withdrawn from the funds. Second, real property can go beyond the traditional commercial properties to include public utilities and toll roads, which are more difficult to value and sell. The purpose of owning such nontraditional properties is to further diversify the portfolio.

In sum, knowing the risk and return profiles of various asset classes and their subclasses, as well as the correlation between them, is the key element in building a

* Leverage means borrowing money to purchase investment instruments so as to increase the return on the investment. Short selling refers to borrowing stock shares and selling them, with the hope of purchasing them back at a lower price in the near future. The objective of short selling is to make a profit even when the stock market declines.

successful investment portfolio that will ensure an adequate long-term rate of return for the pension plan while taking on minimum investment risk.

5.3 Investment Policy

At the center of investment management is the design of an investment policy. Investment policy is important for at least two reasons. First, it clearly lays out the responsibilities of each participant in the investment process. Second, it formalizes and communicates to the outside world, including the plan sponsor, plan members, taxpayers, and the financial market, the governing board's investment goals, objectives, strategies, and policies of the pension system. Once it is designed, all aspects of investment management should be guided by the investment policy.

In general, an investment policy is intended to provide both specific guidelines as well as certain flexibility in asset management, and usually includes the following key elements:

- Investment objectives and risk tolerance
- List of permissible and impermissible asset classes and investment strategies to control risk
- The system's long-term strategic asset allocation, including specific targets and ranges
- The roles and responsibilities of the essential parties, including the governing board, the staff, the investment consultants, investment managers, and custodian(s)
- Standards and measures of investment performance, including benchmarks for each asset class and the system as a whole
- Portfolio rebalancing process

What follows is a discussion of these aspects of an investment policy.

5.3.1 Investment Objective and Risk

An investment objective is a critical element of the investment policy, as it sets the tone for the all-important asset allocation strategy. Investment objectives should grow out of the pension system's investment horizon and the need to meet future pension benefit payments. As the investment horizon is very long-term by nature for a pension system, the overall investment objective of a pension system, therefore, should reflect this long-term horizon. The financial need of a pension system is captured by the system's assumed rate of return. Because the assumed rate of return plays such an important role in determining how fast the asset value should grow in order to meet future pension benefits, it becomes a natural starting point for the investment objective.

An investment objective can usually be articulated in two ways: absolute return and relative return. The absolute return objective means that the investment return of the pension plan has to reach a minimum threshold, which naturally is the plan's assumed rate of return. Since the average assumed return is 8 percent, this means that for most public pension plans 8 percent is the minimum return objective. As the assumed rate of return is a long-term average return, the investment policy should also indicate over how long a time period this minimum return should be achieved. Therefore, it is likely for an investment policy to state that this return should be achieved over a five-year period or over an entire business cycle. The relative return objective means that the actual return has to match or exceed some kind of market index. Pension investment returns are tied to two kinds of indices. The first is the inflation index. As inflation erodes the purchasing power of assets, especially over the long run, achieving a return that exceeds the inflation rate by a certain percentage indicates a real rate of return for the pension plan. The second index that is tied to the return is a customized market benchmark.* The purpose of tying to a benchmark is to ensure the pension plan's return does not lag behind the overall financial market's return based on the plan's asset allocation. Simply comparing fund performance to actuarial rate of return fails to reflect the return the fund could have achieved. For example, if a pension fund returned an average of 8.5 percent over a five-year period, but the underlying customized market benchmark averaged 10 percent over the same five-year period, the pension fund had underperformed even though it had beaten the pension plan's assumed rate of return of eight percent. Conversely, a pension plan return could fall short of the actuarial rate of return, but outperform the markets.

Because these various investment objectives address different concerns facing a pension plan's investment, therefore, it is fairly common for a public pension plan to have multiple investment objectives in the investment policy. For example, the investment policy for the Connecticut state retirement system includes the following investment objectives (State of Connecticut Treasurer's Office, 2002):

- A nominal rate of return equal to or greater than the actuarially assumed investment return of 8.5 percent over the length of a market cycle (generally, a three- to five-year period)
- Achieving a real rate of return of 550 basis points over and above inflation as measured by the Consumer Price Index

---

* A customized benchmark is a weighted average return from a basket of policy benchmarks adopted to measure the performance of the investment portfolio, with the weight being the percentage of portfolio allocated to one particular asset class. For more on policy benchmarks, please see the section 5.3.5 Performance Measurement later in this chapter.

- Realizing returns consistent with or in excess of specific market benchmarks over the length of a market cycle at the individual asset class level
- Maximizing returns in order to minimize long-term contributions within reasonable and prudent levels of risk to the extent investment results can so impact contribution levels

As these investment objectives cannot be achieved without taking some risk, the investment policy also needs to discuss the level of risk the pension system is willing to endure to achieve the return objectives, even if just in fairly general terms. There are several factors that influence the risk tolerance level. First, is the time horizon. The longer the investment time horizon, the more likely an investor can wait out the short-term or even intermediate-term volatility and, thus, more risk can be taken. The long-term nature of pension fund investment should become an important factor in deciding the level of risk to take. The second factor is demographics, i.e., the ratio of active members versus retirees. When the ratio is high, most of the assets will not be needed for a long time and there will be plenty of pension contributions coming in to pay for the current retirees' benefits. Assets, therefore, can be invested for a much longer period and more risk can be taken. When the ratio turns low, more assets are needed in the near future to pay for benefits and, thus, the investment horizon becomes somewhat shorter.

5.3.2 Legal List to Control Risk

Many pension plans also have a legal list of permissible investment instruments and strategies to further control investment risk. The legal list of permissible instruments specifies what types of assets are allowed for investment and more importantly what types of assets are not allowed to be purchased. For example, three states, West Virginia, Indiana, and South Carolina did not permit investment in equity until the late 1990s. Indiana voters passed a referendum in 1996 to allow its public pension systems to invest in equity. South Carolina voters passed a constitutional amendment in 1996 to allow public pension systems to invest in equity, and West Virginia votes did so in 1997. South Carolina voters passed another constitutional amendment in 2006 to allow its pension systems to invest for the first time in international equity. While all states now allow for equity investment, it is still fairly common for states not to permit alternative investments, or some types of alternative investments.

In addition, the amount of some assets that can be purchased can also be limited to a certain percentage of the entire portfolio. This usually applies to the equity portion of the portfolio. Many public pension plans set an upper limit of allocation to equity. For example, when equity investment was finally permitted in West Virginia, the maximum limit for equity allocation was set at 60 percent. Georgia limits its investment in international equity to 10 percent of its portfolio. Other common investment restrictions include limiting the maximum amount of assets that can be placed in one company, in foreign stocks or bonds, or in real estate.

The second part of the legal list has to do with the investment strategies that are not permitted. The two investment strategies that are most likely to be prohibited are short selling and leveraging. The purpose of short selling is to make a profit even when the overall stock market is down. The risk of such a strategy is the explicit bet made on the downward movement of price. If the stock price actually increases, the short-seller will stand to take a loss. Leveraging means borrowing funds to purchase securities. In this case, the explicit bet made through such a strategy is that the stock price will go up. Both the return and the loss to the portfolio will be magnified if the bet turns out to be correct or incorrect, thus adding more volatility to the portfolio. For example, the Indiana Public Employees' Retirement Fund (2007) prohibits short sale of any kind and buying and selling on the margin.

5.3.3 Asset Allocation

Asset allocation is the process of diversifying an investment portfolio among several asset classes (stocks, bonds, real estate, etc.) in order to achieve investment objectives while controlling risk. If the design of investment policy is at the center of investment management, then asset allocation is at the heart of the investment policy. In a 1986 study, Gary Brinson, Randolph Hood, and Gilbert Beebower (1986) examined to what extent portfolio return is attributed to (1) asset allocation policy, (2) marketing timing, and (3) security selection. After analyzing the data on 91 large pension funds, the authors found that 94 percent of the portfolio return can be attributed to asset allocation. This finding was confirmed in a similar study in 1991 on 82 large pension funds (Brinson, Singer, and Beebower, 1991).

Establishing an appropriate asset allocation involves two steps. In the first step, an efficient frontier is developed for the pension plan. In the second step, an optimal portfolio is identified. Mean variance optimization (MVO) continues to be the most common approach used by institutional investors in arriving at an efficient frontier. The following inputs need to be developed to perform an efficient frontier analysis: expected return for each asset class, expected asset class risk (i.e., standard deviation), and the correlation coefficient between the returns of each pair of asset classes in the portfolio. The expected return and risk are based on the historical return and volatility of each asset class, thus giving rise to the name of this approach, with mean referring to average historical return and variance referring to standard deviation, which is the square root of variance. As discussed earlier, a portfolio is considered "efficient" when, compared to all other possible combinations of permissible assets, it produces the highest expected return for a given level of expected risk (or, conversely, the lowest level of risk given a desired level of expected return). A generic MVO analysis with all possible asset classes, however, may not meet the requirement of a public pension plan. It also needs to be tailored to the unique circumstances of a particular pension system, such as prohibition against a certain asset class like alternative investment or limit on the amount of assets that can be allocated to a certain class. In other words, the efficient frontier can be different for different pension systems, depending on the restrictions on investment set by the pension system.

In the second step, the optimal portfolio is identified. The optimal portfolio is the efficient portfolio that best matches the investment objective and risk tolerance of the pension system. This optimal portfolio tells the pension system what asset classes are to be invested in and the percentage of the portfolio devoted to each asset class. This optimal portfolio forms the basis for the pension system's asset allocation strategy. This strategy can be defined both broadly and narrowly. When defined broadly, only the major asset classes and their percentages are identified, such as equity and fixed income. When defined more narrowly, all the subclasses and their percentages are also identified. These percentages associated with the asset classes and subclasses are the pension plan's policy targets, or long-term strategic asset allocation targets. The actual asset allocation, however, may be different from the long-term strategic asset allocation targets, mostly as a result of the changes in the value of financial assets over time. When the value of an asset is increased relative to other assets, that asset becomes overweight relative to the original asset allocation target for that asset. One or more of the remaining assets, thus, have to be underweight relative to the overall policy benchmark.

Due to such drift and the potential risk it introduces to the portfolio, the asset allocation strategy also typically sets a range or band around these policy targets, meaning that the actual asset allocation for any asset class or subasset class has to stay within this band. If an asset class or subclass exceeds the range, the portfolio needs to be rebalanced to bring it within the range. The narrower the range, the more frequently this needs to be done. Table 5.1 shows an example of Arizona State Retirement System's (ASRS) strategic target asset allocation and permitted ranges.

Table 5.1 Arizona State Retirement System Asset Allocation Target

Asset Class	*Policy Target (%)*	*Low (%)*	*High (%)*
Equity			
U.S. equity	53	48	58
International	15	40	20
Fixed income	20	21	31
Real estate	6	4	8

Source: ASRS Comprehensive Annual Financial Report 2006.

5.3.4 Roles and Responsibilities

The main parties involved in the management of pension asset investment are the board of trustees, staff (primarily the chief investment officer and the investment unit), investment consultants, investment managers, and custodians.

5.3.4.1 Responsibilities of the Board of Trustees

While the board of trustees has full responsibility for investment management in most public pension systems, for some pension systems, however, this responsibility lies somewhere else. There are two other institutional arrangements for this important function. The first and fairly similar arrangement is to set up another investment board separate from the retirement system's governing board. This investment board is also governed by a board of trustees and its only responsibility is to manage investment. For example:

- In New Jersey, the State Investment Councils, with eleven members appointed by the governor and the boards of five pension funds, has the overall investment responsibility. The actual management is performed by the Division of Pensions and Benefits within the Department of Treasurer, under the supervision of the Investment Council.
- The Washington State Investment Board consists of fifteen members and manages investments for all state retirement systems in Washington.
- The State of Wisconsin Investment Board (SWIB) manages investments for the Wisconsin Retirement System. SWIB, governed by a nine-member board of trustees, sets the overall investment policy.

In the second and more unique arrangement, a government official is the sole fiduciary of the public pension assets in four states: New York state comptroller and the state treasurers in Connecticut, Michigan, and North Carolina. As a sole fiduciary of pension assets, this official alone is responsible for investment management. Of these four officials, only Michigan's state treasurer is appointed by the governor with Senate confirmation. The other three are elected.

Regardless of the institutional arrangement, whether it is the governing board of the pension system or an independent investment board or a government official, anyone who is responsible for pension asset investment has to abide by the standard of fiduciary responsibility. For the sake of simplicity, anyone who has full responsibility for investment management is referred to as "the board" in this chapter.

The standard of fiduciary responsibilities has shifted over time with the advancement of modern investment management theory. The earliest standard of fiduciary responsibility was enshrined in the "prudent man" rule, which requires the exercise of judgment and care that persons of prudence, discretion, and intelligence exercise in managing their own affairs, not in regard to speculation, but in regard to the income to be derived from there as well as the safety of their capital. By the time ERISA was signed into law, the standard had shifted to the "prudent investor" rule, which was described in the previous chapter. While most public pension systems have adopted the "prudent investor" or "prudent expert" rule, "prudent person" rule is still used by some public pension systems, as shown in a survey by the Office of New York State Comptroller (2006). The key difference between the prudent expert rule and the prudent person rule is the basis for the comparison. The prudent person rule compares the actions of a fiduciary with those of a rational investor managing his own investments with discretion and commonly understood knowledge of investments. The prudent expert rule compares the actions of a fiduciary with those of others acting in a like capacity. The comparison is, in other words, with an investment expert. The standard of prudence for an investment expert is applied to any investment as part of the total portfolio, rather than to individual investments. The "prudent expert" rule incorporates the portfolio diversification theory. It is the overall risk and return profile of the portfolio rather than that of each part of the portfolios that should be of concern to the fiduciary, and the tradeoff between risk and return is the fiduciary's central consideration.

The board's main responsibility in the investment management process is to set the investment policy, while the actual execution of the policy is left to the investment professionals under the board's supervision. The specific responsibilities of the board relating to the investment management of the plan's assets include:

- Establishing an investment policy covering the main issues discussed in this section, like investment objective and permissible investment instruments
- Establishing the asset allocation strategy for the plan, with advice from investment consultants and internal investment staff
- Selecting qualified investment professionals, including investment managers, investment consultants, and custodians to manage the investment, with advice from internal investment staff and investment consultants
- Evaluating the performance of investment managers and the entire portfolio, with advice from internal investment staff and investment consultants, to make sure policy guidelines are adhered to and the investment objective is being achieved

5.3.4.2 Responsibilities of Internal Investment Staff

The internal investment staff, led by the chief investment officer, primarily provides an advisory service to the board. Their main responsibilities include:

- Assisting the board in developing the overall asset allocation of the system's portfolio

- Participating in the hiring of external investment consultants and money managers based on a policy approved by the board
- Monitoring external managers

In some large public pension systems, the internal investment staff also manages the investment of a portion of the portfolio in order to reduce investment cost.

5.3.4.3 Responsibilities of Investment Consultant

The investment consultant's duty is to work with the board and internal investment staff to manage the investment process in an advisory capacity. Specific responsibilities of the investment consultant include:

- Assisting the board in the development and periodic review of the investment policy, including the asset allocation strategy
- Conducting investment manager searches
- Monitoring the performance of investment managers and the entire investment portfolio to provide the board with the information to determine the progress toward the achievement of investment objectives

Investment consultants carry out these duties in a fiduciary capacity, due to the importance of their recommendations to the long-term investment performance of the pension plan. Therefore, they also have to abide by the fiduciary standard.

5.3.4.4 Responsibilities of Investment Managers

Investment managers are hired by the pension system to manage all or a portion of the pension assets. Each manager specializes in one particular asset class or subclass. The manager will have discretion over how to invest the assets under his management, as long as such discretion is within the policies, guidelines, and limits set by the investment policy. The duties of an investment manager include:

- Buying and selling individual securities within an asset class or subclass
- Periodically reporting investment performance
- Communicating to the pension board any major changes that may have an impact on future investment performance, such as a change in investment strategy and a major change in the investment management organization

Due to the direct involvement in asset management, investment managers also have to acknowledge in writing their fiduciary responsibility to a pension plan. As the selection of investment managers will have an impact on the overall performance of the investment portfolio, it is important to have a selection process that is clearly outlined in writing as to who has the ultimate responsibility and how the managers will be evaluated. The policy should also discuss on what basis the service of an investment manager can be terminated.

5.3.4.5 Responsibilities of Custodian

The custodian maintains physical possession of securities owned by the pension plan, collects dividend and interest payments, redeems maturing securities, and receives and delivers securities following purchases and sales. The custodian also provides a periodic accounting report of all assets owned, purchased, or sold, as well as movement of assets into and out of the pension system accounts.

Unlike the service of investment managers, which can be changed with relative ease, the service of a custodian is far more time-consuming to set up for a large pension plan and is rarely changed once this has occurred, making the initial selection of custodians all the more important in terms of the services provided and their cost.

5.3.5 Performance Measurement

Performance measurement is a critical element because ultimately it is the investment performance that determines the adequacy of the funding level.

Investment performance is measured at many different levels. It is measured at the individual investment manager level, subclass level, asset class level and total portfolio level. Whether at the individual manager level or the portfolio level, the actual investment return is calculated and then measured against the returns of a benchmark and peer group. Returns on benchmarks and peer groups provide objective standards to assist in evaluating a manager's or a total portfolio's investment performance.

5.3.5.1 Calculating Investment Return

As there are multiple cash inflows and outflows during an investment period due to pension contributions and divesture of pension assets for pension benefit payment or pension contribution withdrawals, the timing and magnitude of cash flows have to be taken into account when calculating the investment rate of return. There are two main methods of measuring the investment return over a period with multiple cash flows: dollar-weighted and time-weighted rates of return. Dollar-weighted rate of return is the rate that when applied to the market value of the portfolio at the beginning of the investment period, and all subsequent cash flows, the sum of the future value will be equal to the market value of the portfolio at the end of the investment period. The dollar-weighted rate of return is calculated through an iterative process until the correct rate can be found. The rate found in this way is also called the internal rate of return.

For the time-weighted rate of return, the periodic return at the time of a new cash flow has to be known. The annual return is linked to the periodic return in the following formula:

$$\text{Annual Return} = [(1 + PR^1) \times (1 + PR^2) \times \ldots \times (1 + PR^n)] - 1$$

where PR is periodic return measured in percentage.

When a new cash flow is made, the market value of the portfolio of the day prior to the cash flow is calculated, which is then used to calculate the return in the period between the date of last periodic return calculation and the date of new cash flow. The product of the periodic returns is the annual return for the year.

The measurement of mutual fund performance is done in a similar way. In the mutual fund industry, market value and return are calculated on a daily basis in terms of the price of unit share of the mutual fund. Any change in the market value due to investment return is reflected in the change in the price of the unit share and any increase in the market value due to cash inflow is reflected in the change in the number of unit shares of the mutual fund. The annual return of the fund is calculated by comparing the difference between the beginning price and the ending price of the unit share of the fund.

The dollar-weighted and time-weighted rates of return will be equal if there are no cash flows during the investment period or the periodic returns throughout the entire period remain constant. As investment managers have little control over the timing and size of cash flow, the time-weighted rate of return becomes a better measurement of their performance as it excludes the effect of the timing of investment and withdrawal on the return over the entire period.

5.3.5.2 Measuring Investment Managers' Performances

While the measurement of performance of individual investment managers is important, it is not a meaningful measurement just by itself. That return alone does not indicate the quality of his/her performance. Only by comparing it with returns on investments in a similar asset class or subclass can a judgment be made as to whether the manager's performance is adequate or not. For individual managers, there are two ways to measure their relative performance: against a market benchmark and against a peer group.

5.3.5.2.1 Benchmark Return

As an investment manager is selected to manage assets in a particular asset class or subclass, his performance can be measured relatively easily against benchmark returns for that asset class or subclass. A benchmark usually consists of most, if not all, of the securities in the same asset class or subclass. A fund that holds all such similar securities is called an index fund. The return on an index fund is the weighted average return of all the securities in that fund, with the weight being the market value of each security. The index fund return shows the performance of an entire asset class without any actual management involved, as it does not entail buying or selling of any individual security in the index fund. The purpose of comparison with a market index is to see how much additional value the manager has added to the investment portfolio through active management. There is a benchmark for all major asset classes and subclasses.

For the U.S. equity market, two major indices for the broad market are the Wilshire 5000 and Russell 3000. The Wilshire 5000 includes all publicly traded stocks in the United States, numbering over 7000, whereas the Russell 3000 represents the largest 3000 stocks traded publicly. There are also many other indices for the subclasses of the domestic stock market depending on capitalization and style. The Standard & Poor's (S&P) 500 index is the benchmark for large cap stocks, whereas the S&P 400 index fund is a benchmark for mid cap stocks. For small cap stocks, the Russell 2000 index is a widely used benchmark. There are also separate market indices for growth stocks and value stocks. To further dissect the equity market, there are also indices based on both capitalization and style, such as lap cap growth stocks and small cap value stocks.

As for international equity indices, the oldest and probably the most popular index is the MSCI EAFE index maintained by Morgan Stanley Capital International. EAFE stands for Europe, Australia, and Far East. It consists of almost 1000 stocks in these geographic areas. A broader international equity index is the MSCI All Country World, ex-U.S. Stock Index. As for emerging market equities, there is the MSCI Emerging Markets Index.

For the U.S. fixed-income securities market, two widely used indices are the Lehman Brothers Aggregate Index and the Lehman Brothers U.S. Universal Index. The Aggregate Index covers only investment-grade securities, whereas the U.S. Universal Index represents securities that are both investment grade and below investment grade. For noninvestment grade securities, Lehman Brothers also maintains a separate index called the U.S. Corporate High-Yield Index. For the global fixed-income market, Lehman Brothers maintains the Global Aggregate Index, Global High-Yield Index, and Global Emerging Market Index.

For the real estate asset class, a popular index used is the NCREIF Property Index. NCREIF stands for National Council of Real Estate Investment Fiduciaries, a nonprofit organization founded in 1982 to produce research on commercial real estate investment. In 2003, this index consisted of close to 4000 properties with a market value of $127 billion (Fisher, 2005).

For the alternative investment asset class, there are two indices maintained by Cambridge Associates LLC for private equity investment: the U.S. Private Equity Index and the U.S. Venture Capital Index. The U.S. Private Equity Index is based on return data compiled on funds representing over 70 percent of the total dollars raised by U.S. leveraged buyout, subordinated debt, and special situation managers since 1986. The U.S. Venture Capital Index is based on return data compiled on funds representing over 80 percent of the total dollars raised by U.S. venture capital managers since 1981. As for hedge funds, Hedge Fund Research Inc. (HFRI) publishes two comprehensive indices, the HFRI Fund Weighted Composite Index, including over 1800 hedge funds, and the HFRI Fund of Funds Composite Index, including over 800 funds of funds.

5.3.5.2.2 Peer Group Comparison

The second way to measure the relative performance of investment managers is to compare their performance to that of their peer groups, consisting of many other managers in the same asset class and subclass. Many financial consulting firms regularly calculate the peer group return profile and some large financial firms also maintain databases of investment performance of fund managers. Morning Star has probably the best known five-star rating system. Once the managers of the peer group have been identified, they are divided into different percentiles based on their performance. A manager's performance is below average if he is in the lower 50th percentile. Such peer group comparison is not just for the immediate past year's performance, but also for annualized returns over the past three and five years. It is important to have multiyear comparisons because one year is not long enough to judge the performance of an investment manager against his peer group.

5.3.5.2.3 Compliance

In addition to measuring the investment performance of the manager, his/her compliance with investment guidelines provided by the pension system also needs to be closely monitored. The purpose of such monitoring is to ensure that the manager's investment holdings and strategies do not lie outside of the permitted securities and strategies for this particular account prescribed in the investment guideline. There is the potential that in order to boost the investment return, a manager may go beyond the permitted securities and purchase securities of a different risk profile. This will alter the risk and return profile of the overall portfolio, which has been carefully constructed.

5.3.5.3 Measuring Total Portfolio Performance

When the returns of all asset classes and subclasses are known, the return of the entire investment portfolio can be easily calculated. Just as in the case of individual investment managers, this return on the entire portfolio should also be compared to a benchmark return as well as to that of a peer group.

Table 5.2 Tucson Supplemental Retirement System Asset Allocation Target

Asset Class	*Target (%)*
Equity	
Large cap	41
Small/mid cap	11
International	15
Fixed income	28
Real estate	5

Source: Tucson Supplemental Retirement System CAFR 2004.

For the total fund, there are two benchmarks: total fund policy index and asset allocation index. Published market indices are weighted to create a "policy index," with the weights equal to the fund's long-term target allocation for each asset class. The policy index serves as an objective measure of the market return for the entire portfolio. For example, the City of Tucson Board of Trustees has developed the following asset allocation guideline, as seen in Table 5.2.

The plan's policy index, closely matching its asset allocation guideline, is based on: S&P 500 Stock Index (41 percent weight), Russell 2000 Stock Index (11 percent), MSCI All Country World, ex-U.S. Stock Index (15 percent), Lehman Brothers Aggregate Bond Index (25 percent), Lehman Brothers High Yield Bond Index (3 percent), and NCREIF Real Estate Index (5 percent).

An asset allocation index is also constructed using published market benchmarks. In contrast to the policy index, the asset allocation index's weights change to reflect the actual asset allocation of the portfolio as it "drifts" away from the policy targets or as tactical decisions are made to overweight or underweight an asset class. Therefore, any difference between the policy index return and asset allocation index return reflects the difference in return as a result of the actual asset allocation's drift from the policy targets.

The peer group for a pension plan's portfolio is the universe of public pension plans. Since no two public pension plans have exactly the same asset allocation strategy due to such factors as demographics, investment objectives, legal restriction, and risk tolerance, comparing with the returns of other public pension plans is not the same as comparing one individual manager's return with that of his peer group. Such comparison, however, still serves a meaningful purpose. If a plan's overall return is significantly below the average return of public pension plans for a period of time, the plan's board then needs to analyze the causes for such discrepancy to see if this is mainly due to a conservative asset allocation strategy or due to investment performance. If it is due to asset allocation, the board needs to further examine if such asset allocation is in line with other pension plans that are substantially similar in terms of investment objectives and plan characteristics.

If necessary, such analysis may lead to change in the investment policy and asset allocation strategy.

5.3.6 Rebalancing

As discussed earlier, due to fluctuation in the financial market, the increase or decrease in the value of assets can cause the allocation for some asset classes to go beyond the range allowed in the investment policy. The chief drawback of such "drift" is that the overall risk level may be higher than allowed by the investment policy. Portfolio rebalancing is necessary to bring the asset allocation back within the range or to the long-term policy targets established in the investment policy. In addition to bringing down the risk level of the overall portfolio, rebalancing has another added potential advantage of increasing the return if done in a consistent way. Rebalancing means selling assets in overweight asset classes and buying assets in underweight classes. Selling overweight assets, typically as a result of increase in valuation, locks in the positive return and buying underweight assets leads to a better return in the long run.* Over time, disciplined rebalancing can enhance performance and reduce the overall risk of the portfolio.

This benefit of rebalancing is offset to some extent by its drawback. Rebalancing entails transaction costs due to buying and selling of financial assets. The benefit of rebalancing thus needs to be weighed against the cost incurred, i.e., controlling the risk level of the overall portfolio versus the frequency of rebalancing. The investment policy should describe the timing and degree of rebalancing. In terms of timing, there are two basic rebalancing policies, one systematic and the other ad hoc. A systematic policy requires that the portfolio be rebalanced periodically, such as quarterly, semiannually, or annually, even if the allocation for each asset class does not exceed the range allowed in the investment policy. An ad hoc policy requires that the portfolio be rebalanced only when an asset class exceeds its range. Compared to the ad hoc policy, the systematic policy may lead to more frequent rebalancing and, thus, greater cost, but also a tighter control over risk. In terms of the degree of rebalancing, the policy should also state whether or not the asset class should be rebalanced to policy target, merely to within the range, or somewhere in between.

5.4 Implementation Strategy

Once the asset allocation strategy is developed, it then needs to be implemented. In implementing the investment program, there are two main issues that the board

---

* The basic assumption behind this logic is that no asset class as a whole will stay overvalued or undervalued forever, and the return will revert to the historical mean over time.

needs to consider: which part of the portfolio should be managed passively or actively and which part of the portfolio should be managed internally or externally. These two issues also are related.

5.4.1 Passive versus Active Management

Passive management refers to a simple buy-and-hold management strategy. In passive management, the manager creates and maintains a portfolio that closely tracks a market benchmark. Because such a benchmark usually includes most, if not all, of the securities in an asset class, a passive manager does not need to select particular securities within that asset class or subclass. Therefore, passive management is also synonymous with indexing. Changes in the portfolio are made only when the same changes are made in the benchmark.

In contrast, active investment involves selection of securities in an asset class and more frequent buying and selling of securities. Through fundamental research and analysis of securities, an active manager seeks to build a portfolio with selective securities in an asset class that will provide a rate of return in excess of the return on a corresponding market benchmark. Such excess return can be achieved through two means. First, by doing fundamental analysis, the manager hopes to find securities that are undervalued compared to others in the same class and, thus, earn an above average return. Second, by employing market timing strategy, the manager hopes to buy securities when the market is down and sell securities when the market is up, as opposed to a buy-and-hold strategy for passive management.

The comparison between the returns of passive and active management also needs to take into account the management cost. There are two types of cost in investment management. First is the cost for the manager. The cost for a passive manager is much lower than for an active manager because there is very little a passive manager needs to do to construct and maintain a portfolio. Second is the transaction cost. This cost is also lower for passive management as it involves minimum trading once the initial portfolio is constructed. Active management involves more trading and, therefore, incurs more transaction cost. An active manager can achieve an excess return over an index fund return only to the extent that such excess return exceeds the extra cost.

The theoretical underpinning of passive management is the efficient market theory. The theory, whose name was first created by Eugene F. Fama (1965) in an article entitled "Random Walks in Stock Market Prices," posits that the financial market is efficient in the sense that securities are fairly valued and their values will be quickly readjusted to reflect any new information available to investors. For the market to be efficient, two conditions need to be met. First, all information that affects the value of securities comes out in a random fashion and cannot be predicted *a priori*. Second, there are numerous buyers and sellers of securities at any given time who constantly adjust the price of the securities based on the information available. Since all information comes out in a random fashion and the price is quickly adjusted to the new information, it is very difficult, if not impossible for investors to find securities that are undervalued and also to time the market correctly by buying low and selling high, in order to earn an excess return. Even if someone can time the market correctly once in a while, it will be very difficult for him to do it consistently over a long period of time. If undervalued securities are difficult to find and timing the market is also hard to do (the two sources of excess return for active management), then it is not possible for active management to consistently earn a return that is higher than the overall market return. Even if such excess return can be achieved, it may also be offset by the higher cost of active management. What is left then is simply buying a basket of securities rather than picking individual securities and holding them until the time they need to be sold. In this way, the investor will earn a market rate of return, which is the essence of passive management or indexing. One of the strongest advocates of passive investment is Burton Malkiel (2000), the author of the classic book, *A Random Walk Down Wall Street* (W.W. Norton & Company).

The argument for active management is that the financial market is not efficient in the sense that securities are not always fairly valued. While many research studies have been conducted to find the inefficiency of the market, one key criticism against the theory centers on the assumption of rationality on the part of investors. For the securities to be fairly priced, investors have to react rationally when adjusting the price based on available information. Investors, however, do not always behave rationally. More often they exhibit a herd behavior, which means buying together when the price goes up and selling together when the price goes down, without analyzing the intrinsic value of securities. This is best captured by economist Robert J. Shiller's irrational exuberance theory (Shiller, 2000). Because investors act irrationally and emotionally, that leads to market inefficiency and creates the opportunity to identify undervalued securities.

Regardless of the theoretical argument for and against the efficiency of the market, the best way to test which management style results in better return is to compare the actual performance of the two styles. Based on the research done by Lipper, a mutual fund research firm, over a 20-year period ending December 31, 2003, the S&P 500 Index, averaging 12.78 percent annually, outperformed 90 percent of the large cap equity funds, averaging 10.54 percent annually (Malkiel, 2005). The difference in return is primarily due to the cost differential. This shows that over a long period of time, it is very difficult for active management to outperform overall market return, at least for U.S. large cap stocks, when cost differential is taken into consideration.

5.4.2 Internal versus External Management

The second implementation issue is the extent to which the investment portfolio is managed internally or externally. Internal management means that the portfolio is managed by an in-house investment team, i.e., the pension system's own employees. External management means that pension assets are managed by professional investment teams hired from outside the pension system. The key factors in determining whether assets should be managed internally or externally are cost, the level of expertise required, and values added.

In terms of cost, a cost benefit analysis can be done to see which method costs less. Internal management can create the opportunity for cost reduction as external management fees, based on a percentage of assets under management, can be greater than the compensation to the pension system's internal employees. However, this potential advantage can be offset by other extra costs in internal management. There is a significant fixed cost to investment management, such as the information technology needed for trading and support staff for functions like internal control, accounting, and performance reporting. Therefore, the size of the overall investment portfolio matters. The fixed cost will be much less when it is spread over a larger portfolio. A large public pension plan is more likely to manage part of the assets internally than a small one.

The second and third considerations are the level of expertise needed to manage the portfolio as well as the additional return such expertise can generate. These two considerations are closely tied to the first implementation issue, i.e., passive or active management. The question here to ask is to what extent the investment manager is expected to exceed the return of a market index. For certain asset classes, such as the U.S. equity and fixed-income securities like U.S. Treasury bonds, in which the market is considered efficient and above market return is difficult to achieve consistently, such assets are more likely to be managed passively. Since passive investment does not require security analysis and selection and the level of specialized expertise is not that high in constructing a passively managed portfolio, the portfolio is also more likely to be managed internally to reduce cost.

As for asset classes in which the market is not as efficient, such as international equity and alternative investments, there are more opportunities for discovering undervalued securities and active management is warranted. Active management requires more intense research efforts and specialized skills than passive management. In extreme cases, such research effort may even involve trips to foreign countries in the case of international stocks and bonds. The specialized knowledge and skills required make it more cost effective to have such asset classes handled by external managers who specialize in these asset classes. As a practical matter, it is very difficult for public pension plans to internally hire and retain a professional investment manager with a high level of specialized expertise, as it is not possible for a public sector organization to compete with the private sector for such expertise in terms of compensation.

There is substantial variation in the portion of public pension plan assets managed internally. On the one side, until recent years, New Jersey had managed all of its pension assets internally. On the other end of the spectrum, all assets except the cash portion (accounting for 2 percent of total assets in 2005) in the portfolio of the Pennsylvania State Employees' Retirement System (2006) were managed by external managers. In between, most pension systems have parts of their pension assets managed internally. The vast majority of assets managed internally are in the U.S. equity and fixed-income asset classes, the asset classes in which the market is believed to be fairly efficient. For example, the New York State and Local Retirement System (2005) had $126 billion in pension assets in 2005. The domestic equity portfolio, about half of the total assets, consisted mostly of S&P 500 and S&P 400 Index funds that were managed internally by the pension system staff. The system's internal staff also managed the long-term and short-term fixed income portfolios that consisted mostly of U.S. government guaranteed securities. At the same time, the international equity portfolio, the real estate portfolio, and the alternative investment portfolio were managed externally.

The New York state case provides an example of the linkage between these two implementation strategies. It suggests that at least some public systems believe domestic equity and bond markets are fairly efficient and no additional value can be gained through hiring external managers. Index funds managed internally are the best way to achieve the desired return at the lowest cost in these markets. At the same time, this example also suggests that public pension systems do not believe the international equity market and alternative investment asset classes are as efficient and, thus, greater value can be added through actively managing such asset classes within a portfolio.

Another reason that a higher percentage of domestic equity asset class is invested passively is that this is typically the largest asset class in an investment portfolio for a public pension system. When the assets of public pension systems grow over time, more and more public pension assets are invested in the U.S. equity market. It, therefore, becomes increasingly difficult to find undervalued securities in such a market, as public pension systems account for an increasing share of this market. Passive investment, thus, becomes the default investment strategy. This is different for the international equity, as it accounts for a much smaller, although growing, share of an average investment portfolio among public pension systems.

New Jersey provides a particularly interesting case in studying the decision on internal versus external management. Until 2005, New Jersey was the largest public pension system whose pension assets were managed exclusively by internal staff due to state policy. In fiscal year 2005, it had close to $70 billion in pension assets. About 80 percent of the assets were invested in domestic stocks, bonds, and cash equivalent, and another 16 percent in international equity (State of New Jersey, 2005). Due to the internal management, New Jersey pension system had no investment in the alternative investment category. The New Jersey State Investment Council adopted a policy in November 2004 to allocate 13 percent of New Jersey's pension portfolio to alternative investments, including real estate, private equity, and hedge funds, to be reached over a period of five to seven years (State of New Jersey, 1/20/2005). The purpose of this policy change in asset allocation was to diversify away from the traditional stocks and bonds. To partly accommodate this substantial increase in allocation of assets to alternative investments, in July 2006, the State Investment Council officials approved a policy to shift roughly a quarter of the state's pension fund from the control of state employees to private money managers within the next year (McNichol, 2006).

5.4.3 Summary

In all, the overarching objective during the implementation stage is to achieve the best return at the lowest cost. Reducing cost is especially important for public pension systems as they grow larger and larger. Even a small reduction in cost can yield significant savings. For example, for many pension systems with $100 billion or more in assets, a mere 10 basis point reduction in management cost means a savings of $100 million, which can be used to reduce employer contribution so that more funds will be available for public service provision.

References

Brinson, Gary P., L. Randolph Hood, and Gilbert L. Beebower. 1986. Determinants of portfolio performance. *Financial Analysts Journal* 42(4): 39–44.

Brinson, Gary P., Brian D. Singer, and Gilbert L. Beebower, 1991. Determinants of portfolio performance II: An update. *Financial Analysts Journal* 47(3): 40–48.

Center for International Securities and Derivative Markets. 2006. *The Benefits of Private Equity, 2006 Update.* http://cisdm.som.umass.edu/research/pdffiles/benefitsofprivateinvestment.pdf (Accessed July 25, 2007.)

Fama, Eugene F. 1965. Random walks in stock market prices. *Financial Analysts Journal* 21(5): 55–59.

Fisher, Jeffrey D. 2005. U.S. commercial real estate indices: the NCREIF property index. Bank for International Settlements. BIS Papers No. 21. http://www.bis.org/publ/bppdf/bispap21zc.pdf (Accessed July 27, 2007.)

Ibbotson Associates. 2006. *Stocks, Bonds, Bills and Inflation (SBBI) 2006 Yearbook.* Chicago: Ibbotson Associates.

Indiana Public Employees' Retirement Fund. 2007. *Restatement of Investment Policy,* last restated on February 9, 2007. http://www.in.gov/perf/pdf/agency/perf_investment_policy.pdf (Accessed March 6, 2007.)

Malkiel, Burton. 2000. *Random Walk Down Wall Street,* 7th ed. New York: W. W. Norton & Company.

Malkiel, Burton G. 2005. Reflections on the efficient market hypothesis: 30 years later. *The Financial Review* 40(1): 1–9.

McNichol, Dunstan. 7/21/2006. "State Acts to 'Privatize' Pension Funds — Officials Say Move To Independent Managers Will Mean Less Reliance On The Stock Market." *The Star-Ledger*, Newark, NJ, July 21, 2006.

Office of New York State Comptroller. 2006. *Prudent Investor Overview.* (It is provided to the author by the Office of New York State Comptroller.)

New York State and Local Retirement System. 2005. *Comprehensive Annual Financial Report, for Fiscal Year Ended March 31, 2005.* http://www.osc.state.ny.us/retire/word_and_pdf_documents/publications/cafr/cafr_05.pdf (Accessed August 7, 2006.)

Pennsylvania State Employees' Retirement System. 2006. *Comprehensive Annual Financial Report for the year ending December 31, 2005.* http://www.sers.state.pa.us/sers/lib/sers/publications/2005_CAFR.pdf (Accessed June 3, 2007)

Securities and Exchange Commission. 2003. *Staff Report: Implication of the Growth of Hedge Fund.* http://www.sec.gov/news/studies/hedgefunds0903.pdf (Accessed October 6, 2006.)

Shiller, Robert J. 2000. *Irrational Exuberance.* Princeton, NJ: Princeton University Press.

State of Connecticut Treasurer's Office. 2002. *Investment Policy Statement for the State of Connecticut Retirement Plans & Trust Funds.* http://www.state.ct.us/ott/pensiondocs/IPS.pdf (Accessed on February 26, 2007.)

State of New Jersey, Division of Pensions and Benefits. 2005. *50th Comprehensive Annual Financial Report,* for the fiscal year ended June 30, 2005. http://www.state.nj.us/treasury/pensions/annrprts_archive.htm (Accessed June 1, 2007.)

State of New Jersey, Office of the State Treasurer. 1/20/2005. News Release: *Investment Council Adopts Policy and Procedures for Alternative Program.* http://www.state.nj.us/treasury/news/2005/p50120a.htm (Accessed July 21, 2007.)

Chapter 6

Pension Plan Management and the Operating Budget

While the discussion of pension plan management up to this point is limited only to the pension plan, in the real world of public finance, a public pension plan does not exist by itself. The funding of public pension benefits is an integral part of financial resource allocation in the public sector. In order to get a more thorough understanding of the management of public pension plans, the discussion needs to be broadened to include the larger world of public budgeting. Public budgeting is a political process that determines what public services should be funded, how much funding should be given to each public service, and the amount of taxes that is needed to fund all these public services. It is by far the most important component of government financial management.

The relationship between public pension plans and public budgeting can be seen most clearly in the government fund structure. All government revenue and spending can be divided into three fund groups, as shown in Table 6.1.

For state and local governments, most of the pension contributions going into the pension trust fund originally come out of the general fund, which is the government's main operating fund. The general fund collects all the major general taxes, such as corporate and personal income taxes, sales tax, and property tax, to fund most of the essential government services, such as education, public safety, healthcare, and welfare as well as the salaries and benefits for public employees. When viewed in this broader context, public pension funding is no longer an isolated

Table 6.1 State and Local Government Fund Structure

Governmental Fund Group	*Proprietary Fund Group*	*Fiduciary Fund Group*
General fund Special revenue fund Capital project fund Debt service fund Permanent fund	Enterprise fund Internal service fund	Pension trust fund Agency fund

pension financing issue, but rather part of the overall resource allocation decision in the public sector. As such a decision is inherently political due to the competing demands for limited financial resources, the decision on public pension contribution cannot escape the political influence. In this chapter, we examine pension plan management in the larger context of budgeting in the public sector. We look at how public budgeting decisions affect pension contribution and how pension management can also have a major impact on public budgeting. Cases on pension plan management in five states are discussed to illustrate the relationship between public budgeting and public pension plan management.

6.1 Underfunded Pension Plans

The relationship between public budgeting and public pension plan management can be best understood in the context of the funding ratio of a pension plan. A pension plan is underfunded (or overfunded) when the funding ratio is below (or above) 100 percent. There are many factors that cause the underfunding to occur, as mentioned in Chapter 3, such as granting credit to services provided prior to the setup of the pension plan, actuarial assumptions not being met in an unfavorable way, and unfavorable investment returns that leads to asset growth slower than liability growth. While these factors are related to pension management itself, there are two other factors that are related to public budgeting and the politics of public budgeting: inadequate pension contribution and pension benefit increase. Inadequate pension contribution funding means that the actuarially determined pension contribution amount is not fully paid by the plan sponsor. Because the actuarially determined pension contribution is meant to keep the pension plan fully funded or to move it toward full funding, a partial payment will leave an originally fully-funded pension plan underfunded, or an originally underfunded pension plan with an ever lower funding ratio. Granting pension benefit increase to past services can also lead to pension underfunding as no pension contributions were set aside in the past to pay for such higher benefits. When these two factors are combined together, i.e., increasing pension benefit and underfunding pension contribution at the same time, the negative impact on pension funding ratio can be quite severe.

6.1.1 Fiscal Illusion and Fiscal Stress

Why do inadequate pension contribution and pension benefit increase occur even if that can lead to pension underfunding or worsen a pension plan that is already underfunded? Two theories, the fiscal illusion and the fiscal stress theories, have been used most often to explain the pension underfunding issue due to these two factors. Fiscal illusion theory suggests that a fiscal illusion is created when the cost of providing public services is perceived to be lower than it actually is. The result of this fiscal illusion is that either the level of public service is provided at a higher level than it should be or the tax burden is lower than it should be. This fiscal illusion happens because part of the cost has been shifted to others or into the future. In the case of pension contribution, because the actuarially determined pension contribution is for pension benefit accrual attributable to the employee service in that period, it, thus, becomes part of the cost of providing the public service in that period. Not fully funding the pension contribution essentially reduces the cost of service provision and shifts part of the cost to the future.

Bleakney (1973) found that public employers could increase retirement benefits without paying for them at the present time in order to increase overall consumption for current services. Bahl and Jump (1974) noticed that retirement expenditures are increasingly relative to total payroll outlays, implying that governments have been more willing to grant increases in retirement benefits than similar increases in wages and salaries. Inman (1981, 1982) argued that pension underfunding comes about as highly mobile taxpayers apply pressure to underfund local pensions. It is an implicit subsidy from future to current taxpayers for the purchase of public services and, thus, stimulates an over-provision of local public goods. Hall and Hovey (1980) also observed that pension costs have been growing at a much faster rate than other types of public expenditures, largely because it is the nature of pensions that one can incur the liability well in advance of incurring the cost. Using 1978 data for state-administered pension plans, Marks, Raman, and Wilson (1988) found that public pension underfunding is related to unionism, which gives employees the political and bargaining power that can lead to higher benefit levels and, thus, unfunded pension liabilities. Using data on 42 pension plans in 1989, Mitchell and Smith (1994) found that greater unionization is associated with lower levels of actual contributions in the public sector. Johnson (1997) noted that the benefits in the public sector are more generous than those in the private sector, due to the government's ability to underfund public pension plans and there is a positive correlation between unionization and pension benefit increase. Sneed and Sneed (1997) found that unfunded pension liabilities allow governments to hide the true cost of public services, leading to a higher level of public spending.

Fiscal stress theory maintains that when a government faces budgetary difficulties, especially during an economic recession that tends to cause severe fiscal stress, one of the methods used to relieve pressure on the operating budget is to reduce pension contributions to free up funds for other more urgent financial needs.

Mitchell and Smith (1994) found that economic distress, a proxy for budgetary stress, causes public employers to contribute less than is required to their pension plans. Using state-level pension data for 1994 and 1995, Chaney, Copley, and Stone (2002) found that the greater the fiscal stress a state faces, the greater the pension underfunding level, and that the existence of balanced budget requirements also leads to greater underfunding. Eaton and Nofsinger (2004) also found that governments facing tight financial constraints are more likely to reduce pension contribution through pension plan manipulation and their pension plans are also more likely to be underfunded.

The common theme for fiscal illusion and fiscal stress theories is that both pension contribution and pension benefit levels are affected by the political decisions concerning the level of service to provide and the level of tax burden to impose on taxpayers. When there is no immediate negative consequence by postponing pension contribution and/or increasing pension benefits in lieu of salary increase, difficult political choices can be avoided to satisfy public employees, taxpayers, and the public in general.

6.1.2 How to Deal with Pension Underfunding

Whether due to actuarial or political reasons, pension underfunding cannot be ignored indefinitely. First, a severely underfunded pension plan, especially one with a growing funding gap raises concern about the government's financial condition that can lead to lower credit rating and higher borrowing cost. Second and more importantly, the funding level can be reduced to such a level that there will not be sufficient assets in the fund to pay for current pension benefits, eventually leading to a drastic increase in pension contribution from the operating budget to cover the shortage.

There are two methods of closing the funding gap. One method is straightforward and already discussed in Chapter 3, namely amortizing the unfunded liability over a period of time, much like paying off a mortgage loan. The advantage of amortization is that the pension contribution increase will be fairly gradual if the amortization cost is a level percentage of payroll, putting less pressure on the operating budget. The disadvantage is that it will also take the pension plan a long time to reach full funding level if the gap is fairly large to begin with.

The second method is to close the gap through government borrowing in the form of pension obligation bonds (POBs). The proceeds of POBs are used to pay off part or all of the unfunded pension liability to the pension system and then the government issuer of POBs pays debt service to the investors in POBs. Compared to the first amortization method, the potential advantages and disadvantages of a POB as well as the impact on the operating budget are more difficult to analyze as it depends on many factors over the life of a POB.

6.1.2.1 Benefits of POB

There are two main advantages of a POB. First, by bringing the funding ratio to 100 percent or close to that at once, it provides a guarantee to plan members about their future benefits. The second advantage of a POB is interest cost savings. This is similar to the benefit of a refunding bond in the municipal bond market. When the interest rate falls below the rate at which an existing bond was issued, the government issuer can issue a new refunding bond to retire the existing bond and then pay off the new bond at the lower interest rate. Similarly a pension plan sponsor can also issue a POB to achieve interest cost savings. As discussed in Chapter 3, when there is an unfunded accrued actuarial liability (UAAL), it will be amortized over a period of time, typically 30 years, with the interest rate equal to the plan's assumed rate of return. If the overall interest rate falls and the plan sponsor can issue a POB at an interest rate below the assumed rate of return, net of any issuance cost, the plan sponsor can issue the POB to retire the UAAL and pay off the POB at a lower interest rate. The similarity between a refunding bond and a POB, however, ends here. The difference between the two makes a POB more complicated to analyze.

6.1.2.1 Risks of POB

At the time of issuing a refunding bond, the expected present value of future interest cost savings is exactly the same as the actual savings realized by the time the refunding bond is paid off. The expected present value of savings at the time of issuing a POB, in all likelihood, will be different from the actual savings realized over the life of a POB. It will be either larger or smaller, depending on the actual rather than the assumed long-term investment return. It is not possible to know with certainty in advance whether the POB will produce long-term savings. This uncertainty is the main risk of POBs. A POB is an arbitrage between the interest rate on the POB and the investment return on the POB proceeds.* After the POB is sold, the proceeds are put in the pension trust fund to be invested together with other funds in the system, based on the system's asset allocation strategy. The investment return on the bond proceeds has to be greater than the interest rate on the POB in order for the POB to produce savings for the plan sponsor. The greater the difference between the two rates, the greater the savings to the sponsor. If the investment return comes in lower than the interest rate, the POB will actually cost the sponsor more than if the POB were not issued. How likely is it that this is going to happen? At the time of issuance, the investment return is expected to be equal to the assumed rate of return over the life of the POB, which is usually 20

* Because of this arbitrage issue, a POB is issued as a taxable municipal bond rather than as a tax-exempt one, even though most POBs are secured by the unconditional general obligation pledge of the government. In some cases, they can also be secured by a plan sponsor's annual appropriation pledge.

to 30 years. The previous question therefore becomes: How likely is the pension system to achieve its assumed rate of return in the long run? If the assumed rate of return is arrived at after careful deliberation of all possible factors, then it should be reasonable to expect that the assumed rate of return can be achieved over a long period of time. If the actual investment return comes in below the assumed rate of return over a long period of time, then the pension system will have a much bigger problem to deal with, as the lower return will affect the entire system assets, not just the POB proceeds. As a matter of fact, even if the investment return fails to meet the assumed rate of return, a POB can still lead to cost savings as long as the actual return stays above the interest rate on the POB, which is a lower threshold for the pension system to meet.

The second potential risk of the POB is that it may introduce more short-term volatility into pension funding ratio and pension contribution. Although over a long period of time the actual return should be expected to be higher than the interest rate on the POB as long as the assumed rate of return is reasonable, it does not necessarily mean that the actual return will be higher than the interest rate every year. It is entirely possible that in some years, the actual return can fall below (and sometimes significantly below) the interest rate on the POB. If the plan becomes fully funded after the issuance of a POB, this plan will be underfunded again, with the unfunded liability greater than if the POB was not issued. Total debt service, including that on the POB and amortization of the new UAAL, will now be higher than if the POB had not been sold. On the other end of the investment spectrum, the short-term actual investment return can also be significantly higher than the assumed rate of return. This will render the plan in an overfunded status and can lead to a reduction in pension contribution. Such volatility in funding ratio and pension contribution needs to be taken into consideration when using POB to eliminate UAAL. Because of such volatility, it is difficult to judge whether a POB leads to cost savings in the short run. A few years of above average return can make a POB look like a great success whereas a few years of below average return can make it look like a failure. With a POB, its ultimate success or failure can only be judged over its entire life.

The third potential risk is closely linked to the second risk of short-term volatility. This has to do with the timing of investment. In normal circumstances, because pension contributions come into the pension system gradually, they are invested gradually over time and, therefore, the timing of investment is not critical. The unfavorable and favorable investment timing tends to offset each other over time. With POB proceeds investment, however, the timing of investment becomes more critical because the bond proceeds are typically much larger than periodic pension contribution and account for a greater percentage of total pension assets. Such large infusion of cash for immediate investment presents a more challenging investment decision. A sharp drop in the financial market after the POB proceeds are invested will lead to a much greater loss and may take a much longer time for the pension plan to recover. Such concern of timing risk, however, is of less importance when viewed in the long-term perspective, if the plan expects to earn the assumed rate of return in the long run. One method to reduce the timing risk is to invest the bond proceeds gradually over a period of time, a strategy also called "dollar cost averaging."

The final potential risk is a reduction in budgetary flexibility. While unfunded liability and POB are both debt to the government, the former is consider a "soft debt," whereas the latter is considered a "hard debt." Unfunded liability to the pension plan is a soft debt due to the plan sponsor's ability to delay pension payment. This creates more budgetary flexibility for the plan sponsor in the situation of fiscal stress as it can free up funds for more pressing program needs, with the assumption that this delay can be made up for in the future. When a UAAL is turned into a POB, the debt service on a POB can no longer be delayed no matter what the financial situation the government issuer is in. Any delay will constitute a default. From a long-term perspective, however, this short-term disadvantage of a POB turns out to be its advantage. While a POB may deprive the plan sponsor of financial flexibility, it also imposes fiscal discipline on the plan sponsor. It is possible that the financial flexibility may cause the unfunded liability in the first place. Any further delay can only make the funding situation even worse. The fiscal discipline imposed by the debt service will force the plan sponsor to make some difficult fiscal choices with regard to spending and revenue.

Other than the risks inherent in a POB, the structuring of debt service payment on a POB is an important aspect to make sure that it is used prudently. A POB can be structured in many ways. It can be structured so that the debt service will be very low in the beginning and the bulk of the debt will not be paid off until many years later. Such a structure is called front-loading interest cost savings, and the main purpose is to provide short-term budgetary relief to the plan sponsor. By delaying the debt service payment, it can lead to higher cost in the future when compared to the amortization cost of pension liability in the absence of a POB. It can also be structured so that the payment is level for the duration of the bond. Level debt payment spreads out the cost savings in the future and prevents the plan sponsor from manipulating the debt service payment for short-term budgetary relief.

6.2 Overfunded Pension Plan

Unlike underfunding, which can be caused by many factors, the main reason that overfunding occurs is that investment return is higher than the assumed rate of return and the pension liability growth. If the pension plan is already at or close to the full funding level, higher investment return in one or several years can elevate the pension funding level over 100 percent. While overfunding is a much better position for a pension plan to be in than underfunding, dealing with overfunding is nonetheless fraught with risks and uncertainties if the pension surplus is used to reduce/offset pension contribution in the near future or permanently increase pension benefits.

1. **Decrease pension contribution**: Just like a UAAL that is amortized over a number of years, a funding surplus can also be amortized over a period of time. The difference is that the amortization cost of a UAAL is added to that year's normal cost, whereas the amortization cost of a funding surplus is subtracted from that year's normal cost. Therefore, it is normal to see a decrease in annual pension contributions from the previous year's level when the pension plan is overfunded. The critical issue is how fast the surplus should be amortized. A fast amortization schedule can substantially reduce the current year's pension contribution, whereas a longer amortization period will lead to a more gradual and smaller decrease in annual pension contribution. In the extreme case, the surplus may be used to offset the entire pension contribution in a year, leading to a "pension holiday."
2. **Increase pension benefit**: If elected officials can increase pension benefits when the plan is underfunded, it is even more tempting to do so when the plan is overfunded. With the surplus, an increase in pension benefits may not lead to any immediate increase in annual pension contribution, as the increase in pension liability can be offset by the pension surplus.

Both decrease in pension contribution and increase in pension benefit contain risk. This risk results from the long-term nature of investment return. The assumed rate of return is a long-term average return. The volatility in the financial market means that in some years the actual return will be higher than the long-term average and in other years it will be lower. The higher returns will be balanced out by lower returns over time. Funding surplus due to higher investment return is then only temporary rather than permanent. It is temporary because the return cannot stay above long-term average forever and will trend lower in the near future and the funding level can dip below 100 percent. By using up temporary surpluses on the permanent benefit increases, elected officials are making an implicit assumption that investment will keep growing at a rate above the assumed rate of return to support the permanent increase. When the below average return eventually arrives, the permanent increase in benefits will further exacerbate the negative impact of below average return on the plan's funding ratio and also the operating budget. The benefit increase also leads to greater normal cost in the future.

Spending down the temporary surplus through fast amortization will have a similar effect on the operating budget. Although fast amortization is not as risky as benefit increase, it can significantly reduce the annual pension contribution in the short run. When below average return arrives to reduce the funding ratio of the pension plan, required pension contribution will again need to be increased. Such volatility in pension contributions adds strain to the government's operating budget. In the worst-case scenario, fast amortization can be combined with permanent benefit increase.

Pension benefit increase and fast amortization of pension surplus not only add volatility to the government's operating budget, but also magnify the volatility already inherent in the operating budget. To see how they can magnify the operating budget volatility, it is important to understand the cyclical nature of government budgeting. The budgets of state governments and some local governments are heavily influenced by the cyclical economy because the main revenue sources in the operating budget, such as sales tax and corporate and personal income taxes, are all positively related to the economy. When the economy is in an upswing, governments enjoy a surge in revenue collection and budget surplus results. When the economy is in downturn, the revenue collection contracts and budget deficit appears. While the cyclicality of a government budget is mostly a result of the ups and downs in revenue collection, the expenditure side of the budget can also reinforce this volatility. One spending item that can reinforce this volatile nature of the operating budget is pension contribution. The ups and downs of the stock market typically go hand in hand with the ups and downs in the economy. When the economy is in an upturn, the stock market is usually in an upward trend as well and that will increase the value of assets in pension plans as well as the funding ratio. For plans that become overfunded, this can lead to benefit increase and/or pension contribution reduction, as discussed earlier. Because this happens at a time when the government enjoys a surplus in the operating budget, a reduction in pension contributions can further increase the size of budget surplus. When the stock market declines, which usually happens when the economy is also in a downturn, the value of assets in pension plans depreciates, resulting in a funding gap. The funding gap leads to an increase in pension contribution from the operating budget. This happens at a time when the government is most likely to experience a revenue shortfall in the operating budget. The increase in pension contribution further exacerbates the operating budget deficit.

Even though pension overfunding is a good problem to have, how to handle it will have implication not only for the long-term fiscal health of the pension plan itself, but also for the government operating budget. Pension plan management, whether when the plan is underfunded or overfunded, is inextricably linked to the operating budget.

6.3 Case Studies

In this section, we examine five states' experience in pension plan management, with the focus on how the management of pension plans is driven by the conditions of operating budgets and, in turn, how the funding status of pension plans can influence operating budgets. While most of the cases here demonstrate the difficulty of managing public pension plans in the larger political context, the case of Florida also shows that with conservative financial planning, public pension plans can remain financially healthy regardless of the ups and downs in the financial market and add minimum volatility to government operating budgets.

6.3.1 Illinois State Pension Systems

The Illinois state government sponsors five state pension systems. They are: Teachers' Retirement System (TRS), State Employees' Retirement System (SERS), State Universities Retirement System (SURS), Judges' Retirement System (JRS), and General Assembly Retirement System (GARS), with TRS, SERS, and SURS being the largest. Collectively, Illinois state pension systems have the largest unfunded pension liability among public pension systems in the United States. Table 6.2 shows the nine-year trend of total unfunded liability and the funding ratio of the five systems.

The real unfunded liability of the Illinois state government was larger than the roughly $41 billion at the end of fiscal year 2006. The story of Illinois pension systems tells the effect of inadequate pension contributions on the funding ratio and how this, in turn, affects the operating budget in the long run.

6.3.1.1 Underfunding of Pension Contribution

Up until fiscal year 1981, the funding of Illinois pension systems was not based on an actuarially required amount, but rather on a policy somewhat similar to pay-as-you-go — the employer contributions went to pay for current year pension benefits and employee contributions were set aside for investment (State of Illinois, 2007). This policy, inadequate as it was, was abandoned in fiscal year 1982 due to fiscal stress. The contribution in 1982 was reduced from the 1981 level. After 14 years, the contribution level in 1995 was essentially unchanged from the 1981 contribution level, whereas at the same time pension benefit payment increased by about 200 percent. Because of the chronic underfunding of pension contributions over such a long period of time, the unfunded liabilities of the five systems rose to $20.8 billion

Table 6.2 Funding Status of Illinois State Pension Systems (in $Millions)

Fiscal Year	*Actuarial Value of Assets*	*Actuarial Accrued Liability (AAL)*	*Unfunded AAL (UAAL)*	*Funded Ratio (%)*
6/30/1998	37,241.81	51,563.92	14,322.11	72.2
6/30/1999	41,442.44	56,787.68	15,345.24	73.0
6/30/2000	45,949.62	61,518.91	15,569.29	74.7
6/30/2001	42,789.34	67,768.87	24,979.54	63.1
6/30/2002	40,252.59	75,198.15	34,945.56	53.5
6/30/2003	40,721.16	83,825.15	43,103.99	48.6
6/30/2004	54,739.00	89,832.40	35,093.40	60.9
6/30/2005	58,577.94	97,178.99	38,601.06	60.3
6/30/2006	62,341.33	103,073.46	40,732.13	60.5

Source: Illinois State Pension Systems' CAFR, various years.

in fiscal year 1996 from $2.8 billion in fiscal year 1972 (State of Illinois, 1997). Because of this growing unfunded liability, Illinois passed a law in 1995, Public Action 88-593, the purpose of which was to force the state government to increase its pension contribution and the system's funding ratio. The law established a funding ratio objective of 90 percent for the state-funded retirement systems by 2045. To give the state government time to adjust to the higher contribution, the law provided a 15-year phase-in period between 1996 and 2010, during which time the annual pension contribution would increase by a certain percentage.* The increase during the phase-in period was intended to raise the contribution to such a level that beginning in 2011 and through 2045, the pension contribution would simply be a level percent of payroll that would lift the funding ratio of the pension systems to 90 percent at the end of 2045.

Due to the strong stock market returns in the late 1990s, the funding ratio of the five systems increased quickly and reached a peak of about 75 percent in 2000. Between 2001 and 2003, when the stock market declined, the funding ratio deteriorated significantly. The funding ratio dipped below 50 percent in 2003 and the unfunded liability also reached $43 billion. To reduce this ever-increasing unfunded liability, in April 2003, Governor Blagojevich signed House Bill 2660 into law as Public Act 93-0002 (State of Illinois, January 2006). The legislation authorized the state to issue $10 billion in POBs. The bond was scheduled to be repaid over 30 years with an interest rate of 5.05 percent. Because the assumed rate of return of the three major pension systems is 8.5 percent, the issuance of a POB could lead to significant reduction in the unfunded liability and interest cost savings over the next 30 years, if the proceeds of the POB were solely used for the purpose of reducing the unfunded pension liability. The actual use of the proceeds and the design of the debt service schedule on the POB, however, limited the potential savings from such a transaction. Of the $10 billion from the POB, only $7.32 billion was for the purpose of reducing the unfunded liability. About $2 billion was used to cover part of the state's pension contribution in fiscal year 2003 and all of the state's contribution in fiscal year 2004. Another $481 million was used for capitalized interest, meaning borrowing this amount to pay for interest on the entire POB in the first year and, thus, delaying the impact of the debt service on the state's budget. Debt service was structured in such a way that the principal payment was backloaded, therefore, there was no principal payment for the first four years, 10 percent of the principal was paid back in the first 15 years, and close to half of the principal was paid back in the last five years (State of Illinois, August 2006). By backloading the principal payment into the final years, the state government essentially frontloaded all the potential interest cost savings from such a transaction into the first few years to cover pension contribution obligation. Such a pension holiday,

* Because the pension contribution rate during this phase-in period is below the actuarially required rate, the unfunded pension liability will continue to grow even if the statutory contribution was fully paid every year during this period.

while greatly reducing the short-term pressure on the operating budget from paying for the pension contribution, did not provide any real long-term benefit to the state pension systems themselves. Even though the funding ratio of the five plans jumped to 60 percent in 2004 due to the infusion of POB proceeds, there was no real change in the funding pressure facing the state government when future debt service payment was taken into consideration.

This funding pressure was clearly evident soon after the issue of the POB. On June 1, 2005, Governor Blagojevich signed SB 0027 into law as Public Act 94-0004. The main purpose of this Act was to reduce the state pension contribution amount in FY 2006 and FY 2007 from the level required by Public Act 88-593 passed in 1995. PA 94-0004 set the pension contribution for FY 2006 and FY 2007 at $938 million and $1.37 billion, a reduction of $1.2 billion and $1.1 billion from the original requirement. The Act also required that the pension contribution for 2008 through 2010 be increased from their original level to make up for the shortfall in 2006 and 2007, so that there would be no impact on the funding ratio of pension systems by 2010. Partly due to the inadequate pension contribution, the unfunded liability continued to increase. By the end of FY 2006, the unfunded liability grew to about $41 billion, very similar to the size of unfunded liability prior to the issue of the POB. If taking the $10 billion of the POB into consideration, the real pension debt of the state government was about $50 billion in 2006, resulting in a funding ratio of about 50 percent. When taking out the impact of the POB, the underlying funding situation of the five systems did not really improve between 2003 and 2006.

6.3.1.2 Pension Benefit Increase

While inadequate pension contribution over the years has been a major factor in the growth of unfunded liability, another contributing factor has been the pension benefit increase despite the fact that all systems were underfunded. For example, Public Act 90-0582 improved retirement benefits for TRS members by changing the benefit multiplier for pension benefit accrual for service benefits beginning July 1, 1998. The benefit multiplier for service earned before July 1, 1998, was determined as follows: 1.67 percent for each of the first 10 years, plus 1.9 percent for each of the next 10 years, plus 2.1 percent for each of the next 10 years, plus 2.3 percent for each year over 30 years. The new benefit multiplier is set to a flat rate of 2.2 percent for each year.

Public Act 92-0566 created an Early Retirement Incentive (ERI) program for certain members of SERS and state employees covered by TRS in 2002. The ERI was created to reduce state payroll and, thus, pressure on the state budget. While the ERI led to a reduction in payroll, it also increased the pension systems' pension liability by $2.3 billion (State of Illinois, June 2006). The legislation that created the ERI program also required the state to contribute $70 million to SERS in FY

2004 toward this liability and to amortize the remainder of the ERI liability with a level dollar payment between fiscal years 2005 and 2013 at an 8.5 percent interest rate. Public Act 94-0004, discussed earlier, eliminated the level dollar ERI pension liability amortization schedule and wrapped the ERI liability into the pension system's overall liability and the funding on that would be guided by PA 88-593. Essentially, the original amortization period was extended from FY 2013 to FY 2045. This extension would reduce the pension contribution related to the ERI liability in the short run at the expense of significant increase in total pension contributions in the long run.

Decades of inadequate pension contributions and periodic pension benefit increases have put considerable pressure on the future operating budget. By 2045, pension contribution alone will amount to about $17 billion, and the total pension contribution amount between 1996 and 2045 will be about $310 billion (State of Illinois, January 2006). To fulfill this financial responsibility will take enormous discipline in the long run. The Illinois state legislature has made one step toward such discipline. In June 2006, The Illinois General Assembly passed Public Act 094-0839, creating the Pension Stabilization Fund as a special fund in the state treasury. The purpose of this bill was to pay down the unfunded pension liability faster by contributing more than required when there is an operating budget surplus.* For each fiscal year when the general assembly's appropriations and transfers from the general fund do not exceed 99 percent of the estimated general fund revenues, the comptroller shall transfer from the General Revenue Fund a total amount equal to 0.5 percent of the estimated general fund's revenues to the Pension Stabilization Fund; and if appropriations and transfers from general funds do not exceed 98 percent of the estimated general funds revenues, then 1 percent of the estimated general fund revenues should be transferred to the Pension Stabilization Fund. The Pension Stabilization Fund money could not be used to reduce or replace the state's regularly scheduled pension payments.

6.3.2 New Jersey Pension System

Like Illinois, New Jersey also provides an example of how inadequate pension contributions and benefit increases can have a severe negative impact on the pension system's funding status. New Jersey state government is responsible for seven pension plans: Public Employees' Retirement System (PERS), Teachers' Pension and Annuity Fund (TPAF), Police and Firemen's Retirement System (PFRS), State Police Retirement System (SPRS), Judicial Retirement System (JRS), Consolidated Police and Firemen's Pension Fund (CPFPF), and Prison Officers' Pension Fund (POPF). The two major pension plans are PERS and TPAF and the discussion of

* The complete text of the act can be accessed at the Illinois General Assembly's Web site at http://www.ilga.gov/legislation/publicacts/94/PDF/094-0839.pdf.

Table 6.3 New Jersey Pension System Funding Status (in $Millions)

Fiscal Year	*Actuarial Value of Assets*	*Actuarial Accrued Liability (AAL)*	*Unfunded AAL (UAAL)*	*Funded Ration (%)*
Public Employees' Retirement System				
31-Mar-97	6,987.22	6,606.71	(380.51)	105.8
31-Mar-98	7,600.62	7,155.04	(445.59)	106.2
30-Jun-99	8,879.92	7,823.58	(1,056.34)	113.5
30-Jun-00	9,743.73	8,538.69	(1,205.04)	114.1
30-Jun-01	11,123.82	9,886.46	(1,237.36)	112.5
30-Jun-02	11,073.16	10,760.56	(312.60)	102.9
30-Jun-03	10,829.95	11,942.30	1,112.35	90.7
30-Jun-04	10,693.51	12,620.38	1,926.87	84.7
30-Jun-05	10,631.35	13,432.53	2,801.18	79.1
Teachers' Pension and Annuity Fund (TPAF)				
31-Mar-97	22,045.48	21,224.48	(821.00)	103.9
31-Mar-98	24,478.86	23,484.40	(994.46)	104.2
30-Jun-99	27,457.45	25,546.08	(1,911.37)	107.5
30-Jun-00	30,203.21	27,404.62	(2,798.59)	110.2
30-Jun-01	35,351.38	32,745.36	(2,606.02)	108.0
30-Jun-02	25,148.25	35,146.59	(1.65)	100.0
30-Jun-03	34,651.83	37,383.73	2,731.91	92.7
30-Jun-04	34,633.79	40,447.69	5,813.90	85.6
30-Jun-05	34,789.39	43,967.93	9,178.54	79.1

Source: New Jersey Pension System CAFR for fiscal year ended June 30, 2006.

the New Jersey case mostly centers on these two plans. Table 6.3 shows the funding status trend for these two plans.

6.3.2.1 Reduction in Pension Contribution

Unlike Illinois, these two plans were overfunded for most of the years during this period. However, in just five years, the funding ratio of these two plans plunged from over 110 percent to less than 80 percent. New Jersey's pension story started in the early 1990s. In 1992, then-Governor Florio, facing a $1 billion plus budget deficit due to the lingering effect of the economic slowdown, started the first of a series of major pension plan funding changes. There were three components to this first major change: (1) changing the valuation of plan assets from book value to full market value, (2) increasing the assumed rate of return from 7 to 8.75 percent, and (3) allowing for actuarial valuation of the plan assets by using a multiyear smoothing technique (State of New Jersey, 2005). While these changes were aimed at reducing the state government's pension contribution, two of the changes, full market value and multiyear smoothing, actually brought the actuarial valuation of assets closer to the standards later set by the Government Accounting Standards Board (GASB). The assumed return of 8.75 percent, however, was much higher than the average return of 8 percent. A higher rate of return leads to smaller pension liability and also smaller pension contribution in the short run.

After Christine Todd Whitman was elected governor of New Jersey in 1993, two important pension changes were initiated that resulted in reducing pension contribution. First, Chapter 62 of New Jersey Public Law 1994 changed the actuarial valuation method from the entry age normal (EAN) to the projected unit credit (PUC) method. Even though PUC is an actuarially acceptable method of pension funding, the major consequence of switching from EAN to PUC is the reduction in normal cost in the beginning. As discussed in Chapter 3, PUC leads to slower pension liability accrual in the beginning, but faster accrual in the future, when compared to EAN. This shift to PUC reduced state and local employer pension contributions by $547.4 million and $946.8 million for FY 1994 and FY 1995, respectively (State of New Jersey, 2005).

The third major piece of legislation affecting pension funding was Chapter 115 of Public Law 1997.* This piece of legislation changed pension contribution in three major ways:

1. It changed how the assets of retirement systems were valued. The valuation method was changed from the market-related actuarial value using a five-year smoothing technique to "full-market" value for the valuation period ending March 31, 1996. As discussed in Chapter 3, when the financial market is on an upward trajectory, full market value is usually higher than actuarial value of assets as the former immediately recognizes the most recent price appreciation instead of spreading such appreciation over several years.
2. The full market value of pension assets should also include the POB proceeds of $2.7 billion. Chapter 114, P.L. 1997 authorized New Jersey Economic Development Authority to issue $2.7 billion in a POB. The issuance of a POB was to pay off the unfunded accrued pension liability and to achieve interest cost savings. At the time, New Jersey had an assumed rate of return of 8.75 percent and the POB was issued at an interest rate of 7.5 percent.
3. Third, it would allow employer contribution to be offset by the excess valuation pension assets according to the following schedule: pension contribution can be offset by up to 100 percent of the excess valuation assets for valuation

* The complete text of this legislation can be accessed at the New Jersey Legislature's Web site at http://www.njleg.state.nj.us/9697/Bills/PL97/115_.PDF (Accessed on March 15, 2007.)

periods ending March 31, 1997 through March 31, 2001, and up to 84 percent, 68 percent, and 50 percent of the excess valuation assets for the valuation periods ending March 31, 2002, March 31, 2003, and March 31, 2004, respectively.

These three changes worked together to reduce employer contribution. First, it resulted in an immediate amortization of pension fund surplus, at least for the first few years, by offsetting pension contribution with 100 percent of pension surplus. Second, by using the full market value rather than the actuarial value of assets, the law maximized the excess valuation assets available for offsetting pension contribution. Third, since the POB proceeds were included in the valuation of assets and any excess valuation assets could be used to offset pension contribution, then any offset in pension contribution was made possible because of the borrowed funds rather than actual excess valuation assets. According to Table 6.3, the combined excess valuation assets in the two largest systems did not exceed $2.7 billion in 1997 and 1998, although they exceeded that amount in later years. While the issuance of the POB led to a reduction in interest cost owed to the various pension systems, it was also instrumental in reducing the pension contributions in the following years.

In addition to reducing employer contribution, Chapter 115 also reduced employee contribution. For calendar years 1998 and 1999, the rate of contribution for plan members was reduced by 0.5 percent from excess valuation assets. Thereafter, the rate of contribution of plan members for a calendar year would be reduced equally with excess valuation assets to the extent possible. Chapter 415, P.L. 1999 further reduced the employee contribution rate by another two percentage points for calendar years 2000 and 2001 with excess valuation assets.

6.3.2.2 Pension Benefit Increase

The next major piece of legislation that affected pension funding was Chapter 133, P.L. 2001, which increased the retirement benefits under service, deferred, and early retirement by changing the formula from 1/60 (1.67 percent) to 1/55 (1.82 percent) of final compensation for each year of service.* This legislation also increased the retirement benefit for veteran members with 35 or more years of service and reduced the age qualification from 60 to 55. The law further provides that existing retirees and beneficiaries would also receive a comparable percentage increase in their retirement allowances.

To pay for the additional annual employer normal cost contribution due to the increased benefits, the law establishes a benefit enhancement fund for both TPAF and PERS, which would be funded by excess valuation assets beginning with the valuation period ending June 30, 1999. The law set the valuation of assets for the

* The complete text of this legislation can be accessed at the New Jersey Legislature's Web site at http://www.njleg.state.nj.us/2000/Bills/PL01/133_.PDF (Accessed on March 15, 2007.)

period ending June 30, 1999, to be the full market value of the assets as of that date. It also affirmed the sliding scale of pension contribution offset in Chapter 115, P.L. 1997. By choosing the full market value in 1999 when the stock market was close to the end of the unprecedented run-up in valuation, significant amount of excess pension assets could be used to offset the increase in pension normal cost and the pension benefit increase.

Table 6.4 shows the state pension contribution between 1997 and 2006. The excess employer contribution in 1997 came from the proceeds of the POB rather than from the state's operating budget. When taking this into account, it means that the New Jersey state government avoided paying almost anything into its two large state pension plans over 10 years.

In addition to the pension benefit increase from Chapter 133, P.L 2001, other pieces of legislation were enacted that provided pension enhancements. For example, Chapter 23, P.L. 2002 provided early retirement incentive benefits to state employees who met specified age and service requirements and who retired within a specified time period. The incentive benefits included an additional three years of service credit to employees who are at least 50 years of age with at least 25 years of service credit; and an additional monthly benefit of $500 per month for 24 months following the date of retirement to employees who are at least 60 years of age with at least 10, but not more than 20, years of service credit.

6.3.2.3 Consequence

Because of years of avoiding paying pension contribution by using up the excess pension assets and increasing pension benefits at the same time, the pension funding ratio deteriorated significantly because it had reached its peak. Both the PERS and TPAF peaked at over 110 percent in 2000. This was much higher than the national average ratio of about 100 percent in that year. By 2005, the two systems' funding ratio dropped to about 80 percent, lower than the national average. The 30-percent drop in the funding ratio was more than twice the national average over the same period. This severe drop in funding ratio, therefore, is not mainly due to the financial market. This difference in funding performance can only be attributable to the twin factors of contribution elimination, even after the funding ratio fell below 100 percent in 2003, and benefits increased. This is made worse by using the full-market value of pension assets. By using this method of valuation at the peak of the stock market price in 1999, much of the excess assets were used up, leaving little to smooth out the stock market downturn in the next few years and, thus, contributing to a more precipitous drop in funding ratio compared to other major public pension systems that use a market-smoothing asset valuation method.

The New Jersey case provides a classic example of how liberal usage of pension accounting and a POB to avoid pension contribution obligation and increase pension benefits comes at the expense of long-term fiscal health of both the pension

Table 6.4 New Jersey Pension System Employer Contribution

	Public Employees' Retirement System			*Teachers' Pension and Annuity Fund*		
Fiscal Year	*Annual Required Contribution*	*Employer Contribution*	*Percentage Contributed (%)*	*Annual Required Contribution*	*Employer Contribution*	*Percentage Contributed (%)*
1997	134,878,582	241,106,642	178.80	372,060,546	1,601,688,633	430.50
1998	78,833,287	0	0.00	297,219,462	0	0.00
1999	86,945,810	0	0.00	314,671,482	258,816,649	82.20
2000	103,033,425	0	0.00	368,904,564	0	0.00
2001	85,078,620	0	0.00	0	0	n/a
2002	88,911,187	0	0.00	0	0	n/a
2003	44,636,619	0	0.00	194,453,594	0	0.00
2004	50,365,892	526,505	1.00	686,284,850	0	0.00
2005	115,017,395	463,342	0.40	883,460,483	0	0.00
2006	153,436,981	568,129	0.40	1,177,674,055	93,834,999	8.00

Source: New Jersey Pension System CAFR for fiscal year ended June 30, 2006.

plan and the government operating budget. In its recommendations on reforming the state pension program, the New Jersey Pension Task Force targeted one of these factors by suggesting that any legislative bill to increase pension benefits in the future is required to identify a revenue source and certify its costs and that revenues have been developed in accordance with generally accepted actuarial principles (State of New Jersey, 2005).

6.3.3 New York State and Local Retirement System

The New York State and Local Retirement System, also known as the Common Retirement Fund (CRF), with an actuarial asset value of $140 billion in 2006, is the third largest state pension system in the country. It consists of two pension plans, the Employees' Retirement System (ERS) and the Police and Fire Retirement System (PFRS). As the ERS is far larger than the PFRS, the case of the New York state and local pension system is focused on the ERS only.

CRF is one of a few public pension systems that use the aggregate cost actuarial method and, thus, does not have historical data on the pension system's funding ratio. Due to the lack of data on the funding ratio, annual investment return is used in its place as a proxy for the pension system's funding status and its impact on pension contribution, assuming that a higher return is generally associated with a better funding ratio. Similar to other cases in this chapter, New York is a story of skipping (or attempting to skip) pension contribution during bad economic times and reducing pension contributions during good economic times, as can be seen in Table 6.5. This table shows the state and local government pension contribution rate went almost full circle over a 20-year period from about 10 percent of employee salary at the beginning of the period to almost 0 percent and then back to over 10 percent again at the end of this period.

However, what differentiates New York State from other cases in this chapter is its unique pension governance structure. As discussed in Chapter 4, instead of a board of trustees appointed by the executive and legislative branches of the government, CRF is governed by the state comptroller, an elected official who is the sole trustee of the CRF. This independence from other elected officials has increased the willingness on the part of the pension system to challenge attempts by the state government to skip or reduce pension contribution during economic downturn.

6.3.3.1 Pension Contribution Reduction by Governor

Over this 20-year period, the New York state government attempted to skip or reduce pension contribution several times when the state budget was under pressure. The first major attempt occurred in 1990 when New York was hard hit by a severe economic recession and the resulting large budget deficits. To help solve the budget problem, the state government made deep cuts into pension contribution by

Table 6.5 New York State Pension Contributions (in $Millions)

Year	*Employer Contribution*	*Percentage of Salary (%)*	*Rate of Return*	*Year*	*Employer Contribution*	*Percentage of Salary (%)*	*Rate of Return*
1987	978.4	9.4	17.8	1997	668.8	3.7	10.9
1988	1,102.6	9.7	1.6	1998	280.1	1.7	30.4
1989	573.7	3.7	13.4	1999	193.2	1.3	8.8
1990	284.5	3.6	13.9	2000	83.0	0.9	17.8
1991	–188.2	0.3	11.7	2001	131.0	0.9	–8.7
1992	181.2	0.4	10.7	2002	179.1	1.2	2.8
1993	140.9	0.6	12.5	2003	525.5	1.5	–10.2
1994	337.2	0.7	6.9	2004	1,052.3	5.9	28.8
1995	47.2	0.7	8.8	2005	2,434.5	12.9	8.5
1996	521.8	2.2	21.8	2006	2,271.4	11.3	14.6

Source: New York State and Local Retirement System, CAFR 2006.

passing Chapter 210 of the Laws of 1990. This law changed the actuarial method of the pension system from the aggregate cost method, which had been in use for 70 years up until then, to the PUC method. As discussed earlier, PUC leads to smaller pension liability accrual in the beginning and smaller pension contribution. The pension contribution was reduced to such an extent that in 1991 the state pension system had to refund the state government for overpayment in the past. Chapter 210 of the laws of 1990 was challenged in court, on the grounds that it violated the state Constitution. Section 7 of Article V of the New York State Constitution states: Membership in any pension or retirement system of the state or of a civil division thereof shall be a contractual relationship, the benefits of which shall not be diminished or impaired. The state Supreme Court upheld the challenge in 1993 and affirmed the order of the Court of Appeals. The Supreme Court applied the State Constitution's Nonimpairment Clause in rejecting the state's switch to PUC from aggregate cost:

> Chapter 210 impairs the benefits of the existing pension fund. Said legislation allows employers to deplete monies in the existing pension fund by reducing the amount of employer contributions. Employers are allowed a credit of a portion of the existing monies, and need not contribute to the pension until the reserved monies are drastically reduced. To later replenish the fund, employers and employees must increase the amount of their contributions to the pension fund. As such, the reserve monies will not be available for immediate investment, the return on investment of monies in the existing fund will be significantly decreased, and the additional security provided by the reserve monies in the pension funds will be impaired.*

To avoid an immediate hit to the state budget because of the switch back to the aggregate cost that would increase the pension contribution rate, the state comptroller decided to phase in the rate increase. The rate would increase by 1.5 percent a year starting in 1996 and ending in 1999. This phase-in plan, however, was carried out only in 1996 and 1997 when the rate was increased to 3.7 percent from 0.7 percent in 1995.

The second major attempt by the state government to reduce pension contribution happened at the end of 2003 during preparation for the budget of fiscal year 2004–2005. Amid fiscal pressure on the budget and in anticipation of the higher pension contribution in the new fiscal year because of the severe drop in asset value in the previous three years, the state government proposed a legislative bill with many pension reform proposals. Several of the proposals had the effect of reducing the state government's pension contribution: eliminating or widening 20 percent asset valuation "corridor," limiting annual increase in pension contribution rate

---

* *McDermott v. Regan*, supra, 82 NY2d 354, 604 N.E.2d 985, 604 N.Y.S. 2d 890 (1993).

to two percentage points, and establishing 30-year amortization for 2000 benefit enhancements.* The governor's budget office estimated that the pension reform bill, if passed, would save state and local governments a total of $1.3 billion in pension contribution in the next fiscal year.† However, the Office of State Comptroller considered these proposals unconstitutional, as they would result in substantial reduction in employer pension contribution and, thus, harming the constitutionally guaranteed pension benefits of employees (Office of the New York State Comptroller, 1/9/2004). Due to the legal precedent, this pension bill did not go any farther.

Again for fiscal year 2005–2006, the executive budget proposal appropriated less than the required contribution amount to fund the State's pension obligation. The Comptroller objected to the underfunding. The Legislature concurred and added $169.2 million, bringing the State's appropriation up to $836.4 million, the required contribution amount (Office of the New York State Comptroller, 5/2005).

6.3.3.2 Pension Contribution Reduction by Comptroller

While the state pension system led by the state comptroller was instrumental in preventing the state government from reducing its pension obligation during a time of fiscal stress, it also was responsible for reductions in employer pension contribution several times during this 20-year period, although in a very different environment when the investment return was much higher than expected. The first major decision came in 1989 when the pension system increased the assumed rate of return from 8 percent to 8.75 percent, in response to the strong investment returns over the past few years. This upward shift in the assumed rate of return resulted in a smaller pension liability and was responsible for part of the substantial drop in the contribution rate in 1989 to 3.7 percent from 9.7 percent a year earlier. Had the assumed return not been changed upward, then the state's action to change the valuation method most likely would not have led to a contribution rate as low as 0.3 percent in 1993. This assumed return was revised downward to 8.5 percent in 1997 and farther down to 8 percent in 2001.

The second major pension system decision came in the late 1990s. The extraordinarily strong investment returns between 1996 and 2000, as can be seen in Table 6.5, led to the so-called "pension holidays" in 1998, 1999, 2000, and 2001. As discussed earlier, in the aftermath of the court decision to revert back to aggregate cost method, the pension system would increase the employer pension contribution

* Expansion of the 20% corridor leads to higher asset valuation and, thus, lower pension contribution. The original amortization period for 2000 pension benefits was 17 years.

† New York State Division of Budget. 2004–2005 New York State Executive Budget Public Protection and General Government Article Vii Legislation Memorandum In Support. http://www.budget.state.ny.us/archive/fy0405archive/fy0405articleVIIbills/ppgg_memo.html#prtHbudget (Accessed March 23, 2007.)

rate by 1.5 percent a year. Due to the very strong investment returns, the state comptroller actually decreased the contribution rate to 1.7 percent in 1998 rather than increase it to 5.2 percent based on the phase-in plan. This rate continued to decrease for the next few years until it was below 1 percent in 2001. Such rate reduction happened at a time when state and local governments were awash in budget surpluses due to the strong economy in the mid- to late 1990s.

6.3.3.3 Consequence

Because of the stock market downturn in 2000, 2001, and 2002, the contribution rate had to go up significantly again. In order to ease the pain of sudden dramatic increase in the pension contribution rate on the state and local governments at a time when they were already under fiscal stress, and also to prevent such volatility in contribution rates from happening in the future, State Comptroller Alan Havesi proposed a major pension reform bill. In May 2003, the state legislature passed and the governor signed the pension reform bill, known as Chapter 49 of the Laws of 2003 (Office of the New York State Comptroller, 5/15/2003). This law had two major components in terms of employer pension contribution rate. First, to reduce the volatility in pension contribution rate in the future, the new law requires employers to make a minimum contribution of 4.5 percent of payroll every year, including years in which the investment performance of the fund would make a lower contribution possible. Second, to ease the pain of sudden increase in pension contribution rate in FY 2004–2005, local governments would be permitted to bond any contributions in excess of 7 percent of estimated salaries. In addition, the state comptroller may allow state and local governments to amortize required contribution above 7 percent of salaries over five years. Employers who do so will be required to pay 8 percent interest on the amortized amount. Without these provisions, the required FY 2004 contributions from state and local governments would have been $2.7 billion, rather than the roughly $1.1 billion paid in that year (Snell, 2004).

The temporary budgetary relief provided in Chapter 49 of the Laws of 2003 was extended in Chapter 260 of the Laws of 2004 to further ease the burden on participating governments' operating budget due to the sudden substantial increase in pension contribution (Office of the New York State Comptroller, 8/2004). The new law allowed participating governments to amortize or issue bonds to pay a portion of their pension contribution for three fiscal years: for fiscal year 2005, the amount in excess of 7 percent of salaries is eligible for bonding or amortization; for fiscal year 2006, the amount in excess of 9.5 percent of salaries is eligible for bonding or amortization; and for fiscal year 2007, the amount in excess of 10.5 percent of salaries is eligible for bonding or amortization. Further, the amortization period was set for 10 years rather than 5 years in Chapter 210 of the Laws of 2003. The interest rate on the unpaid amount was also changed to the market interest rate on similar securities rather than the system's assumed interest rate of 8 percent. This rate was set to 5 percent for fiscal years 2005 and 2006. For the fiscal year 2005 pension bill, the amortized amount receivables from New York State and participating employers were $473.2 million and $119 million, respectively, as of March 31, 2006. For the fiscal year 2006 pension bill, the amortized amount receivables from New York State and participating employers were $155.1 million and $32 million, respectively, as of March 31, 2006 (New York State and Local Government Retirement System, 2006).

This brief story of the New York pension system offers two lessons. First the unique pension governance structure limits to some extent the reduction in pension contribution by the state government, which can be best appreciated in comparison with New Jersey. The independence of the New York pension system trustee allows it to be more willing to challenge any government attempt to reduce pension contribution. Second, even such independence cannot render the system impervious to the temptation of pension contribution reduction in times of strong investment return and, thus, the risk of volatility to state and local government operating budgets. The minimum required contribution rate, adopted in 2003, however, adds a strong measure of protection to the pension system and substantially reduces such risk in the future.

6.3.4 West Virginia Retirement System

While Illinois has the largest unfunded liability, the West Virginia state pension system had the lowest pension-funding ratio until 2007. West Virginia has established six state-level pension plans: Teachers Retirement System (TRS), Public Safety Death, Disability, and Retirement Fund (PSDDRF), State Police Retirement System (SPRS), the Judges Retirement System (JRS), Deputy Sheriff Retirement System (DSRS), and Public Employees Retirement System (PERS). The West Virginia state government is largely or fully responsible for the first four plans and TRS is the largest by far, with PSDDEF being a distant second. The discussion of the West Virginia case centers on these two plans. Table 6.6 shows the trend of funding ratio of these two plans. For most of the time during this period, the funding ratios of these two plans hovered around 20 percent. The ratio for PSDDEF, however, jumped to 75 percent in 2005.

6.3.4.1 Underfunding of Pension Contribution

Like Illinois, one of the main reasons that West Virginia had such a low funding ratio was underfunding of pension contribution over a long period of time. Because of its rust-belt economy and the nationwide economic recession, the economic growth was slow in the 1980s and early 1990s, resulting in slow growth in tax revenue and fiscal stress. One target for relieving the fiscal stress was contributing less than the required amount to the pension plan. The underfunding was compounded

Table 6.6 West Virginia Retirement Plan Funding Status (in $Millions)

	Actuarial Value of Assets	*Actuarial Accrued Liability*	*Unfunded AAL (UAAL)*	*Funded Ratio (%)*
TRS				
30-Jun-98	839.6	4,262.8	3,423.2	19.7
30-Jun-99	932.5	4,629.3	3,696.8	20.1
30-Jun-00	1,046.8	4,883.1	3,836.2	21.4
30-Jun-01	1,090.7	5,188.8	4,098.2	21.0
30-Jun-02	1,098.4	5,709.0	4,610.6	19.2
30-Jun-03	1,191.7	6,243.98	5,053.0	19.1
30-Jun-04	1,427.5	6,440.7	5,013.3	22.2
30-Jun-05	1,627.4	6,617.7	4,990.4	24.6
PSDDRF				
30-Jun-98	67.6	297.8	230.3	22.7
30-Jun-99	74.2	343.3	269.1	21.6
30-Jun-00	84.9	361.0	276.1	23.5
30-Jun-01	89.2	388.9	299.7	22.9
30-Jun-02	91.1	416.9	325.8	21.9
30-Jun-03	99.4	447.9	348.5	22.2
30-Jun-04	11.8	462.1	344.0	25.6
30-Jun-05	361.4	485.4	124.0	74.4

Source: West Virginia Retirement System CAFR various years.

by another unique problem facing the West Virginia pension system. As mentioned earlier, the state was one of only three that constitutionally prohibited investment in the equity market until the late 1990s. This forced the state to invest its pension assets in money market funds and fixed-income securities. While much of the volatility in the financial market was avoided with such investment, the long-term growth rate of pension assets was severely limited. Even though West Virginia passed a constitutional amendment in 1997 allowing the state to invest in the equity market, it first bought stocks in February 1998 and its allocation reached 60 percent in early 2000, which is the legal limit. In other words, it missed out almost entirely on the longest running bull market, which started in the early 1980s and ended in 2000, with a few interruptions in between. It was really this long bull market that most other pension systems took advantage of to propel the average funding ratio to 100 percent in 2000. Because of the low long-term return and inadequate pension contribution, West Virginia saw its funding ratio sinking to the bottom of the nation's public pension systems.

In 1994, a West Virginia Supreme Court order to fully fund the state's teachers' retirement plan prompted the state to take action to restore the financial health of its pension systems. The West Virginia state legislature devised a 40-year plan in 1994, similar to the one in Illinois, which called for gradually ramping up the pension contribution level to pay off the unfunded liability. The payment plan compels the legislature to increase West Virginia's pension contributions by about 6.5 percent a year.

6.3.4.2 State Efforts to Reduce Unfunded Liability

Given the significant drain on state financial resources due to the increasing pension contribution, the West Virginia state government attempted to find ways to reduce the long-term cost of paying off the pension debt. In the early 2000s, to take advantage of the low interest rate, then Governor Bob Wise's administration tried to issue a $3.9 billion pension obligation bond (POB), authorized by the Pension Liability Redemption Act in 2000, to address the pension-funding problem. The bond was challenged in court by State Auditor Glen Gainer II and Treasurer John Perdue, who argued that the offering violated the state constitution, which prohibits the state from taking on general obligation bonds without voter approval because the POB would be issued as a general obligation bond. In December 2004, the West Virginia Supreme Court of Appeals ruled unanimously that the POB could not be issued without voter approval. In February 2005, the West Virginia state legislature passed a Pension Bond Amendment Resolution as an amendment to the state Constitution and put the POB on the ballot in that year. By 2005, the total unfunded liability already grew to $5.5 billion. The Pension Bond Amendment would permit the sale of up to $5.5 billion of general obligation bonds to help fund the gap in the four plans the state is largely responsible for, especially the TRS. To make the POB more attractive to both voters and potential investors, West Virginia promised more fiscal discipline in pension management in the future and security to the investors in the POB, in the amendment:

> The bond resolution shall pledge that while any of the bonds are outstanding, the State will take all reasonable actions to prohibit any unfunded actuarial accrued liability from occurring in the pension systems administered by the State. The bond resolution shall further pledge, and the indenture shall so state, that while any of the bonds are outstanding, should any increase of existing benefits or the creation of new benefits under any of the pension systems administered by the State, ... cause any additional unfunded actuarial accrued liability in any of the pension systems administered by the State (calculated in an actuarially sound manner) during any fiscal year, such additional unfunded actuarial accrued liability of that pension system will be fully

Table 6.7 West Virginia Pension Contribution (in $Thousands)

	PSDDRF		*TRS*	
Year	*Required Contribution*	*Percentage Contribution (%)*	*Required Contribution*	*Percentage Contribution (%)*
30-Jun-99	12,591	119.0	202,899	100.7
30-Jun-00	17,545	95.9	217,676	97.5
30-Jun-01	18,105	100.0	232,249	100.2
30-Jun-02	19,742	103.2	237,688	110.2
30-Jun-03	22,473	100.2	271,808	105.4
30-Jun-04	24,701	99.0	307,634	106.8
30-Jun-05	27,953	904.9	342,800	103.0

Source: West Virginia Retirement System CAFR various year,

amortized over no more than the five consecutive fiscal years following the date the increase in benefits or new benefits become effective.*

The voters rejected the bond issue by 54 percent to 45 percent. After the POB failed, the state officials, led by Governor Joe Manchin, resorted to two other means to pare down the unfunded liability. First, a burgeoning economy, spurred largely by surging energy prices, helped the state's revenues grow in fiscal 2005 and produced large budget surpluses. That allowed the state government to use the budget surpluses to make a one-time payment of about $280 million to PSDDRF, as can be seen in Table 6.7. Due to this infusion of excess contribution, the funding ratio of PSDDRF jumped from 25.6 percent in 2004 to about 75 percent in 2005, as indicated in Table 6.6.

After PSDDRF was largely fully funded, West Virginia started tackling the much bigger unfunded liability in TRS. In 2006 and 2007, the West Virginia economy continued to grow very fast and generated large budget surpluses for the state government. In the three fiscal years from 2005 to 2007, the state government appropriated an additional $768 million more than the state was required under the 40-year plan to pay toward the Teachers' Retirement System debt (West Virginia Governor's Office, 6/26/2007). Including the nearly $300 million in additional payments to PSDDRF over the same period, the state government contributed $1 billion more than required to the state pension systems, a significant amount compared to the size of West Virginia's state general fund, which averaged about $3.5 billion during this period.

* The full text of this legislation can be accessed at the West Virginia Legislature's Web site at http://www.legis.state.wv.us/Bill_Text_HTML/2005_SESSIONS/lx/Bills/sjsr101%20eng.htm.

To further increase the funding ratio of TRS, the state government took a second action. The state legislature in 2007 passed SB 185, creating the Tobacco Settlement Finance Authority to sell more than $800 million in tobacco settlement bonds, the proceeds were to pay down the unfunded liability in TRS.* On June 14, 2007, West Virginia received $807 million from the sale of tobacco settlement bonds (West Virginia Governor's Office, 6/14/2007). Between the additional payment of $1 billion and the tobacco bond sale, the annual payment to state pension systems was reduced from an eventual estimate of over $720 million at the end of the 40-year plan to a level $285 million over the remaining years of the original 40-year plan, resulting in a savings of $2.5 billion, and the funding ratio of the Teachers' Retirement System also increased from 22 percent to 51 percent in 2007 (West Virginia Governor's Office, 6/26/2007).

While West Virginia suffered from similar pension woes, its approach to solving the problem was different from that of Illinois. Illinois has been borrowing to pay for pension contribution and continuing to underfund contributions so as to free up funds for other government programs, whereas West Virginia has been using budget surpluses to make additional pension contributions to pay down the unfunded liability.

6.3.5 Florida Retirement System

While all the previous cases show the effect on the long-term fiscal health of pension plans of inadequate and reduced pension contributions as well as pension benefit increases, Florida Retirement System (FRS) is a story of prudence and constraint. The success of FRS can most easily be seen in the trend of funding ratio, as shown in Table 6.8.

With an actuarial value of $117 billion in assets under management and a funding ratio of the 106 percent in 2006, FRS is one of the few large public pension systems that were overfunded in 2006. However, as recently as 1991, Florida's pension system was only 60 percent funded. By 1998, the system was already overfunded. By 2000, its funding ratio reached a peak of 118 percent, much higher than the average of 100 percent for large state pension plans at that time. What is most remarkable about FRS is that its funding ratio remained stable in the years after 2000, whereas many other large pension plans experienced significant drops in pension funding ratio.

6.3.5.1 Steady Asset Growth

How did FRS double its funding ratio in just ten years and also survive the stock market decline with little damage to its funding ratio? Because the ratio is determined by assets and liabilities, both sides of the funding ratio need to be examined

* As the tobacco settlement bond is backed by the funds from the tobacco companies and not the state's general obligation, the sale of such bonds do not require voter approval.

Table 6.8 Florida Retirement System Funding Status (in $Millions)

Fiscal Year	*Actuarial Value of Assets*	*Actuarial Accrued Liability*	*Unfunded AAL (UAAL)*	*Funded Ratio (%)*
1-Jul-91	23,484	39,259	15,775	58.82
1-Jul-93	31,693	45,656	13,963	69.42
1-Jul-95	41,586	54,027	12,452	76.95
1-Jul-97	56,221	61,611	5,390	91.25
1-Jul-98	66,997	63,206	(93,791)	106.00
1-Jul-99	77,795	68,575	(9,220)	113.75
1-Jul-00	88,504	74,949	(13,555)	118.09
1-Jul-01	95,518	80,994	(14,524)	117.93
1-Jul-02	99,406	86,470	(12,936)	114.96
1-Jul-03	101,907	89,251	(12,655)	114.18
1-Jul-04	106,707	95,185	(11,522)	112.10
1-Jul-05	111,540	103,926	(7,614)	107.33
1-Jul-06	117,160	110,978	(6,182)	105.57

Source: Florida Retirement System CAFRs, various years.

for the answer. On the asset side, the system has adopted a fairly aggressive, but also very consistent, investment strategy. It has consistently allocated close to 70 percent of assets in stocks, and the rest for bonds and real estate, as can be seen from Table 6.9.

Table 6.9 Florida Retirement System Asset Allocation

Year	*Equity (%)*	*Bonds (%)*	*REITs (%)*	*Alternative (%)*	*Cash (%)*	*Return (%)*
1996	64.12	25.24	2.91		7.73	17.33
1997	68.45	24.15	3.16		4.24	21.30
1998	69.80	25.05	3.87		1.28	22.00
1999	71.44	23.73	3.83		1.00	14.00
2000	67.62	24.20	3.94	3.40	0.84	10.50
2001	66.06	25.28	4.11	3.58	0.97	–7.60
2002	63.68	26.32	4.25	3.33	2.43	–8.07
2003	67.20	23.00	4.60	3.40	1.40	2.85
2004	70.00	20.00	5.70	3.40	0.50	16.65
2005	70.00	21.70	4090	3.20	0.30	10.18

Source: Florida Retirement System CAFR, various years.

Table 6.10 Florida Retirement System Pension Contribution (in $Millions)

Year	*Required Contribution*	*Percentage Contributed (%)*	*Year*	*Required Contribution*	*Percentage Contributed (%)*
1993	2,529	100	2000	1,969	111
1994	2,734	100	2001	1,869	110
1995	2,880	100	2002	1,825	97
1996	2,967	100	2003	1,844	98
1997	3,036	100	2004	2,044	92
1998	3,184	100	2005	2,141	102
1999	3,102	100	2006	2,193	96

Source: Florida Retirement System CAFR, various years.

In the 10-year period between 1996 and 2005, its investment return reached double digits in seven years. Despite the negative returns in 2001 and 2002, its annualized return over the 10-year period between 1995 and 2005 was 9.37 percent, significantly above its assumed rate of return of 8 percent. Higher than assumed return over a long period of time is the first reason why the funding ratio improved so dramatically in the late 1990s and remained steady since.

While investment return accounts for the majority of the increase in asset value, another important factor in asset accumulation is pension contribution. The system's pension contribution has been fairly steady during this entire period, as can be seen in Table 6.10.

The pension contribution amount increased gradually since the early 1990s until it peaked at $3.18 billion in 1998, a year in which the funding ratio first surpassed 100 percent. When the funding ratio reached its height of 118 percent in 2000 with an actuarial excess value of $14 billion, the pension contribution was reduced by about $1 billion. Such reduction, however, is understandable as the large surplus in the system had to be amortized over time and this amortization led to a reduction in pension contribution. What is most remarkable is that the pension contribution was not reduced further. As a matter of fact, pension contribution remained fairly steady at about $2 billion for the next six years after the initial reduction in 2000, despite the fact that FRS was sitting on large excess assets. This steadiness in pension contribution is due to the establishment of a rate stabilization mechanism by Florida State Legislature in 2000 in Florida Statute Chapter 121.031. The rate stabilization mechanism operates as follows:

1. The actuarial surplus shall be the value of actuarial assets over actuarial liabilities, as is determined on the preceding June 30 or as may be estimated on the preceding December 31.

2. If the actuarial surplus exceeds 5 percent of actuarial liabilities, one-half of the excess may be used to offset total retirement system costs. In addition, if the actuarial surplus exceeds 10 percent of actuarial liabilities, an additional one-fourth of the excess above 10 percent may be used to offset total retirement system costs. In addition, if the actuarial surplus exceeds 15 percent of actuarial liabilities, an additional one-fourth of the excess above 15 percent may be used to offset total retirement system costs.
   Any surplus amounts available to offset total retirement system costs pursuant to the previous paragraph should be amortized each year over a 10-year rolling period on a level-dollar basis.

Two important features of this mechanism provide the stabilizing force. First, it establishes a pension surplus reserve. By allowing only part of the excess assets to be used on pension cost reduction, the rest of the excess assets are essentially put in a de facto reserve fund, even though no such fund actually exists. Minimally, the first five percent of the excess assets cannot be used for contribution reduction. Such reserves then can be used to offset any below average return or other unfavorable actuarial experiences in the future, thus avoiding an increase in contribution rates in such a situation. Second, even the part of the surplus that can be used for contribution reduction has to be spread out over a rolling 10-year period, resulting in a much smaller immediate reduction and more stable contribution over time. It is due to these two features of the rate stabilization mechanism that pension contributions remained so stable since 2000. Had these pension contributions been substantially reduced or even eliminated for all these years because of the pension surplus, as done by other states in previous examples, then the system's funding ratio would have dropped below 100 percent by 2006.

6.3.5.2 Constraint on Pension Benefit Increase

The liability side of the equation also shows signs of constraint. This is partly due to the legal constraint on the state government. In 1976, voters in Florida passed an amendment, Article X of the Florida Constitution, which prohibits any public pension system in the state from increasing benefits without concurrently making provision for the funding of those benefits on a sound actuarial basis (Florida Advisory Council on Intergovernmental Relations, 1979). The Florida state legislature then added Part VII to Chapter 112 of the Florida Statutes to implement this amendment, with the following provisions: (1) the contributions made by the employer and employee must be sufficient to meet the normal cost, plus an amount sufficient to amortize the unfunded liability over a period no longer than 40 years, (2) the employer-provided portion of the retirement benefit, when added to the member's Social Security benefits, shall not exceed 100 percent of the final average compensation. Part VII lays out the legislative intent:

> It is the intent of this act to prohibit the use of any procedure, methodology, or assumptions the effect of which is to transfer to future taxpayers any portion of the costs, which may reasonably have been expected to be paid by the current taxpayers. Actuarial experience may be used to fund additional benefits, provided that the present value of such benefits does not exceed the net actuarial experience accumulated from all sources of gains and losses.

This essentially prohibits any pension benefit increase when the pension plan is underfunded as any such increase will shift the cost to future taxpayers. Because of the constitutional and statutory requirements on pension benefit increase and the upper limit set on pension benefits, there has not been any major pension benefit increase in the Florida Retirement System since the late 1970s, even when the system's funding ratio reached 118 percent in 2000. This can be seen in Table 6.8. The accrued liability has been increasing at a fairly steady pace of about 8 percent each year with no major upward bumps.

By holding steady on pension contribution and pension benefits during the financial market upturn, the Florida system was able to wade through the financial market downturn without any damage. By doing this, it avoided any significant increase in pension contribution in recent years that has added much financial stress to many state and local government budgets. Such conservative pension management practice has made the Florida Retirement System one of the healthiest among large state pension systems. FRS continued to demonstrate such conservatism in other aspects of pension management. In 2004, the system lowered its assumed rate of return from 8 percent to 7.75 percent, putting it below the average assumed rate of return for major pension systems. While this will lead to an increase in accrued pension liability, which can be absorbed more easily when the system was enjoying a healthy surplus, it also makes sure the system will have a better chance of meeting the financial responsibilities in the future when the return may not be as robust as in the past.

6.3.6 Summary

These stories of pension plan management show a consistent theme, which is the tight linkage between pension contribution and the government operating budget. The pension contribution and the funding ratio are susceptible to two economic forces: (1) the economic cycle that affects the operating budget and (2) the investment cycle that affects the value of pension assets. These two forces reinforce each other and create volatility in the level of pension contribution, funding ratio, and government operating budget.

6.4 Pension Contribution Stabilization Methods

As can be inferred from the cases, maintaining pension contribution stability is not only important for the pension systems themselves in terms of long-term fiscal health, it is also important for the government operating budget. Even though some methods have already been discussed in the previous section, more systematic discussion of this subject, which is the main subject of this final section in this chapter, is needed given its importance.

6.4.1 *Fixed Contribution Rates*

As the name of the method suggests, the pension contribution is fixed at a constant percentage of payroll, which is approved by the legislative body that establishes the pension plan. This contribution rate does not change whether the plan is underfunded or overfunded. The major advantage of such a method is that it completely eliminates any volatility in pension contribution and is very predictable year after year from a budgeting perspective. This is especially valuable when the pension plan is overfunded. By maintaining the same level of contribution, the government is paying more than the actuarially required amount, thus building up a pension contribution reserve. When the financial market goes down and the actuarially required contribution level increases due to the deterioration in funding level, the government can contribute less than the required amount, using the reserve to make up for the difference. This will free up some funds in the operating budget for spending on other services, just at a time when the government is facing a revenue shortfall. Several state and local government systems have adopted such a method, such as those in Wyoming, Texas, and Minnesota. For example, the Wyoming Retirement System fixes the contribution rate at 5.57 percent of payroll. Table 6.11 shows the system's funding ratio and its contribution over an eight-year period.

Due to the fixed rate, the actual contribution increased gradually over this period because of an increase in payroll. When the pension system was overfunded, the actual amount was significantly above the required amount, and when the system was underfunded in 2003 and 2004, the actual amount fell below the required amount. When the funding ratio improved in 2005, the actual amount outpaced the required amount again. Over the entire period, the state contributed $40 million more than the required amount. This was achieved without any volatility in the actual contribution and in the operating budget, despite the fact that the required pension contribution increased from $11.7 million in 2001 to $90.5 million in 2004.

One potential disadvantage of such a method is that the actual contribution based on the fixed rate may be significantly below the required contribution when the financial market suffers a major setback, especially for a prolonged period of time. The funding ratio can be further eroded due to inadequate pension contribution when the required rate is substantially above the fixed statutory rate. Therefore,

Table 6.11 Wyoming Public Employee Pension Plan Contribution

Year	*Funding Ratio (%)*	*Required Contribution ($)*	*Actual Contribution ($)*	*Percentage Paid (%)*
1998	102.4	31,602	46,143	146.0
1999	105.8	38,637	48,577	125.7
2000	113.8	27,673	50,487	182.4
2001	103.2	11,700	56,517	483.1
2002	92.2	51,463	57,377	111.5
2003	91.7	82,740	55,364	66.9
2004	85.0	90,477	60,574	66.9
2005	95.1	60,574	65,192	107.6
Total		394,866	440,233	111.5

Source: Wyoming Retirement System CAFR, various years.

to make this method successful in the long run, the fixed rate needs to be linked to the long-term actuarial rate that can fully fund all future pension benefits and needs to be adjusted higher when there is an extended period of time in which the actual investment return is below average. In the case of the Wyoming Retirement System, the funding policy also requires that in the event that the current statutory contribution requirement is not sufficient to accumulate assets to pay pension benefits, the pension board, with approval of the state legislature, would revise the statutory rate (Wyoming Retirement System, 2006).

6.4.2 Minimum Contribution

This method can be viewed as a variation of the previous fixed-rate method. As in the case of the New York State Retirement System, minimum contribution means that the actual contribution cannot fall below a fixed rate approved by the legislative body. Even when the pension plan is overfunded and the required contribution rate is below the minimum rate, the minimum rate has to be met, essentially also building a temporary pension contribution reserve in this situation. While the minimum rate sets a lower limit, it does not set a ceiling on the contribution and, thus, the contribution rate can go substantially higher when the pension plan is underfunded as a result of lower than expected return in the financial market. Compared to the previous method, while the minimum rate method provides some stability in pension contribution, it does not have the same level of stability. However, it does offer one advantage over the previous method, which is an automatically higher contribution rate when the pension plan is underfunded. Just like in the previous method, the success of this method also lies in the determination of the minimum rate. Ideally, this minimum rate can be set to the level that should at least fund the normal cost for each year.

6.4.3 Indexing Contribution Rate to Prior Year's Level

This method is useful when the contribution rate is much lower than the rate in the prior year. By indexing to the prior year's rate, the current year's contribution rate will include both the actuarially determined rate and a portion of the reduction. Rhode Island has such a funding policy for its pension system. Effective July 1, 2005, the state law requires, commencing in fiscal year 2006 and each year thereafter, for each fiscal year in which the actuarially determined state contribution rate for state employees and teachers is lower than that for the prior fiscal year, the governor shall include an appropriation to the pension system equivalent to 20 percent of the rate reduction to be applied to the actuarial accrued liability. The amount to be appropriated shall be included in that annual appropriation bill and paid into the retirement system. The retirement system's actuary can adjust the computation for the annual required contribution for the year in which supplemental contributions are received; such contributions once made may be treated as reducing the actuarial liability remaining for amortization in the next following actuarial valuation. Such rate indexing limits to some extent the volatility in both the pension contribution rate and the operating budget.

6.4.4 Corridor Funding

Corridor funding means that the pension contribution rate, determined on an actuarial basis, will remain constant as long as the pension plan's funding ratio stays in a range. For example, Maryland State Retirement and Pension System (2005) adopted such a method:

- Effective July 1, 2002, when the funding ratio for each of the combined systems is at least 90 percent, but not more than 110 percent, the contribution rate will be the rate for the previous fiscal year.
- If the funding ratio is below 90 percent, the contribution rate will be the sum of the contribution rate for the previous fiscal year and 20 percent of the difference between the full funding rate for the current fiscal year and the contribution rate for the previous fiscal year.
- If the funding ratio is above 110 percent, the contribution rate will be the difference between the contribution rate for the previous fiscal year and 20 percent of the difference between the contribution rate for the previous fiscal year and the full funding rate for the current fiscal year.

The advantage of corridor funding is that the pension contribution rate can remain very stable within a fairly broad range for the funded ratio. The disadvantage is that if the funding ratio falls below 90 percent and stays there for a period of time, then the pension contribution rate under such a method will take much longer for the pension plan to get above the 90 percent ratio than an actuarially determined rate. In 2005, the Corridor Funding Committee, appointed by the Maryland pension system's board, recommended an immediate transition to actuarial funding, as the pension system's funding ratio had fallen below 90 percent and actuarial study found that it would remain below that level for many years. The board of trustees approved the committee's recommendation (Maryland State Retirement and Pension System, 2005).

6.4.5 Pension Fund Surplus Reserve

The purpose of a pension fund surplus reserve is to preclude a certain percentage of pension surplus from being amortized and used to reduce future pension contribution. A pension fund surplus reserve is automatically created with this preclusion. The creation of such a reserve is in recognition of the fact that a pension plan is overfunded largely as a result of higher than expected investment return. Because the investment return will not stay at that level forever, such pension surplus is mostly temporary in nature. When the investment return eventually turns lower, this temporary surplus reserve provides a cushion against any funding gap. The Florida Retirement Systems sets an example in this case. Any pension fund surplus that is less than 5 percent of accrued pension liability has to be set aside and cannot be used for amortization.

6.4.6 Extend Pension Surplus Amortization

This method seeks to minimize the effect of pension fund surplus on future contribution by amortizing the surplus over a long period of time. Since unfunded pension liability is usually amortized over a 30-year period to reduce its impact on pension contribution, then pension surplus should also be amortized over a long period of time to reduce its impact on future pension contribution. As typified in the case of New Jersey, most pension plans do not have specific policies specifying how fast to amortize pension surplus, and this has become a major factor for pension contribution volatility. The Florida Retirement System sets an example in this respect. It amortizes a portion of the pension surplus on a rolling 10-year basis. With a statutory mandate, such an amortization method significantly minimizes the extent to which a temporary pension surplus can offset future pension contribution, leading to more stability over time. The other advantage is that by leaving a significant portion of the surplus in the pension plan, there is more cushion against future financial market downturn that can cause the funding ratio to decrease and contribution to increase.

6.4.7 Extend Asset Smoothing Period

The purpose of asset smoothing is to gradually recognize investment gains and losses and, thus, to stabilize pension contribution. While three to five years is fairly common among public pension systems as the smoothing period, several governments have recently extended their asset smoothing periods. For example, the California Public Employees' Retirement System (2005) has lengthened its asset-smoothing period from 3 years to 15 years. The advantage of extending the smoothing period is that smaller portions of annual investment gains and losses are incorporated in determining the funding ratio and, thus, contribution rate. Because this longer period would likely extend over at least a full economic cycle, market downswings would be mitigated by prior market upswings. In the case of an overfunded pension plan due to investment gains, extending the asset-smoothing period has the similar effect on pension contribution as extending the amortization period of pension surplus, as amortizing pension surplus over a long period of time is the same as recognizing investment gains over a long period of time. In both cases, a large portion of the investment gains is set aside for the future.

The disadvantage of this method is that the smoothed value could become significantly different from the market value, consequently giving conflicting signals about plan funding. In this situation, establishing an asset value corridor would help to address these issues. A corridor limits the difference between actuarial and market values of pension assets. For example, in Washington State, beginning with actuarial studies done after July 1, 2003, changes to plan asset values that vary from the long-term investment rate of return assumption shall be recognized in the actuarial value of assets over a period that varies up to eight years depending on the magnitude of the deviation of each year's investment rate of return relative to the long-term rate of return assumption. Beginning with actuarial studies performed after July 1, 2004, the actuarial value of assets shall be between 70 and 130 percent of the market value of assets as of the valuation date.*

6.4.8 Summary

While all of the methods mentioned above are discussed separately, it does not mean only one should be chosen. In reality, one or two or more can be combined to provide more stability in pension contribution and funding ratio. As in the case of Florida, it combines a long amortization period for surplus with a pension surplus reserve.

Because several of the rate stabilization policies discussed in this section came about after 2000, it shows that many public pension sponsors have learned from the

* Washington State Legislature. Revised Code of Washington 41.45.035, Long-term economic assumptions — Asset value smoothing technique. http://apps.leg.wa.gov/RCW/default.aspx?cite=41.45.035 (Accessed July 20, 2007.)

detrimental effect of pension contribution volatility and this bodes well for public pension plan management in future economic and financial market cycles.

References

Bahl, R.W., and B. Jump. 1974. The budgetary implications of rising employee retirement costs. *National Tax Journal* 27(3): 479–490.

Bleakney, Thomas P. 1973. Problems and issues in public employee retirement systems. *Journal of Risk and Insurance* 40(1): 39–46.

California Public Employees Retirement System. 2005. Press Release: *CalPERS Advances Employer Rate Stabilization Plan*, March 15, 2005.

Chaney, Barbara A., Paul A. Copley, and Mary S. Stone. 2002. The effect of fiscal stress and balanced budget requirements on the funding and measurement of state pension obligations. *Journal of Accounting and Public Policy* 21(4&5): 287–313.

Eaton, Tim V., and John R. Nofsinger. 2004. The effects of financial constraints and political pressure on the management of public pension plans. *Journal of Accounting and Public Policy* 23(2): 161–189.

Florida Advisory Council on Intergovernmental Relations. 1979. *Local Government Retirement System in Florida.* http://www.floridalcir.gov/reports/retire79.pdf (Accessed March 20, 2007.)

Hall, Arden and Hal Hovey. 1980. State and local government retirement systems: problems and prospects. *National Tax Journal* 33(3): 371–380.

Inman, R. 1981. Municipal pension funding: A theory and some evidence: A comment. *Public Choice* 37(1): 179–187.

———1982. Public employee pensions and the local labor budget. *Journal of Public Economics* 19(1): 49–71.

Johnson, Richard W. 1997. Pension underfunding and liberal retirement benefits among state and local government workers. *National Tax Journal* 50(1): 113–42.

Marks, Barry R., K. K. Raman, and Earl R. Wilson. 1988. Toward understanding the determinants of pension underfunding in the public sector. *Journal of Accounting and Public Policy* 7(3): 157–183.

Maryland State Retirement and Pension System. 2005. *Recommendation on Corridor Funding Methodology.* http://www.sra.state.md.us/investments/CorridorFundingReport-24Jan06.pdf (Accessed June 26, 2007.)

———2006. *Comprehensive Annual Financial Report for the year ended June 30, 2006.* http://www.sra.state.md.us/annualreports/2006/CAFR2006.pdf (Accessed May 2, 2007.)

Mitchell, Olivia S. and Robert S. Smith. 1994. Pension funding in the public sector. *The Review of Economics and Statistics* 76(2): 278–290.

New York State and Local Government Retirement System. 2006. *Comprehensive Annual Financial Report 2006.* http://www.osc.state.ny.us/retire/word_and_pdf_documents/publications/cafr/cafr_06.pdf (Accessed April 29, 2007.)

Office of the New York State Comptroller. 5/15/2003. Press release: *Governor, Senate and Assembly Enact Hevesi's Plan to Strengthen Pension Fund and Stabilize Governments' Payments*. http://www.osc.state.ny.us/press/releases/may03/051503.htm (Accessed March 24, 2007.)

———1/9/2004. Press release: *Major Pension Revision Proposals Are Unconstitutional*. http:// www.osc.state.ny.us/press/releases/jan04/011904.htm (Accessed March 24, 2007.)

———8/2004. *Modification of NYS Retirement System's Annual Payment Date and Reserve Provision for Pension Contributions*. http://www.osc.state.ny.us/localgov/pubs/ releases/0804.htm (Accessed March 20, 2007.)

———5/2005. 2005–*2006 Budget Analysis: Review of the Enacted Budget*. http://www.osc. state.ny.us/reports/budget/2005/enactedbudget051105.pdf (Accessed March 24, 2007.)

Sneed, Cynthia A., and John E. Sneed. 1997. Unfunded pension obligations as a source of fiscal illusion for state governments. *Journal of Public Budgeting, Accounting and Financial Management* 9(1): 5–20.

Snell, Ronald. 2004. Pensions and Retirement Plan Enactments in 2003 State Legislatures. http://www.ncsl.org/programs/fiscal/pensun03.htm (Accessed March 23, 2007.)

State of Illinois, Commission on Government Forecasting and Accountability. January 2006. *Report on the 90% Funding Target of Public Act 88-0593*. http://www.ilga.gov/commission/cgfa2006/Upload/Funding_PA_88-0593.pdf (Accessed March 25, 2007.)

———June 2006. *Report on the Cost and Savings of the State Employees' Early Retirement Incentive Program (P.A. 92-0566)*. http://www.ilga.gov/commission/cgfa2006/ Upload/2006_JUNE_ERI_Report.pdf (Accessed March 12, 2007).

———August 2006. *Report on the Financial Condition of the Illinois Public Employee Retirement Systems*. http://www.ilga.gov/commission/cgfa2006/Upload/0806Financial_ Condition_%20PublicEmpl_Retire.pdf (Accessed March 25, 2007.)

State of Illinois, Office of State Comptroller. 1997. Funding state pensions — Where we have been and where we are going. *Fiscal Focus*: January. http://www.apps.ioc.state. il.us/ioc-pdf/ff0197.pdf (Accessed March 11, 200.7.

———2007. Illinois state pensions continue to put pressure on the state budget. *Fiscal Focus*: January/February. http://www.apps.ioc.state.il.us/ioc-pdf/FFWeb0107.pdf (Accessed March 11, 2007.)

State of New Jersey, Benefits Review Task Force. December 2005. *The Report of the Benefits Review Task Force to the Acting Governor Richard Codey*. http://www.njspba.com/ Documents/final_report.pdf (Accessed March 13, 2007.)

West Virginia Governor's Office. 6/14/2007. News Release: *State of West Virginia Nets More Than $800 Million from Tobacco Securitization Completed Today*. http://www.wv.gov/ OffSite.aspx?u=http://www.wvgov.org/ (Accessed July 7, 2007.)

———6/26/2007. News Release: *Tobacco Bond Sale Marks Over $2.5 Billion in Savings and Much Lower Debt Payments in Future*. http://www.wv.gov/OffSite.aspx?u=http:// www.wvgov.org/ (Accessed July 7, 2007.)

Wyoming Retirement System. 2006. *Comprehensive Annual Financial Report for Fiscal Year Ended December 31, 2005*. http://retirement.state.wy.us/pdfs/CAFR.pdf (Accessed June 7, 2007.)

Chapter 7

Defined Contribution Pension Plan

As many of the case studies in the previous chapter have demonstrated, defined benefit (DB) pension plans carry with them substantial risks to the plan sponsors. Because of these risks and other perceived drawbacks of DB plans, there has been an increase in calls for public pension reform, especially after the downturn in the stock market between 2000 and 2002, which led to a sharp increase in pension contributions for many plan sponsors and put heavy burden on the budgets of many state and local governments. Two types of public pension reforms have been called for, one incremental and the other more fundamental. An incremental reform calls for a decrease in pension benefit level for public employees so as to reduce the plan sponsor's pension contribution level and pension liability in the future. Due to the constitutional and statutory protection of participants' pension rights, as discussed in Chapter 4, such reform, if implemented, would not affect the benefit level of current employees, although it would create a new tier of reduced pension benefits for new employees.

A fundamental pension reform, however, calls for discarding the DB plan type altogether and replacing it with a defined contribution (DC) plan type. The gradual increase in the number of DC plans in the private sector over the past several decades and particularly the termination of many large DB plans in recent years also lends more support to such a switch in the public sector. In the most notable example, in California, Governor Arnold Schwarzenegger called for changing California's retirement system, the largest in the country, from the DB to the DC plan type in 2002, primarily as a response to the quickly deteriorating funding status of the California Public Employee Retirement System in the aftermath of stock market decline.

While the incremental reform will be discussed in more detail in the final chapter of this book, it is the fundamental reform of public pension benefits that is the focus of this chapter. Since the profiles of DC programs have grown more prominent in the public sector, an understanding of state and local pension plan management will not be complete without understanding the benefits and risks involved in switching from a DB to a DC plan. In this chapter, we will first compare the advantages and disadvantages of both DB and DC plans. Second, the current status of DC plans in the public sector is examined. Third, a hybrid pension plan is discussed that combines the characteristics of both DB and DC plans. Fourth, issues in implementing a DC plan are addressed.

7.1 Defined Benefit (DB) Plan

While some risks of defined benefit (DB) plans have already been discussed in the previous chapter, a more systematic discussion of the advantages and disadvantages of a DB plan is needed in order to understand the implication of switching to a defined contribution plan. Because DB and DC are completely opposite plan types, the advantages of a DB plan tend to be the disadvantages of a DC plan and vice versa.

7.1.1 Advantages of a Defined Benefit Plan

There are several advantages of DB plans for a plan participant. The biggest advantage is the certainty of pension benefits after retirement. There are three aspects to this certainty. First, he or she knows the amount of annual pension benefits because it is determined by a set formula, as discussed in Chapter 2. As long as he can make a reasonable projection of how long he will work in the public sector and how fast his salary will grow, a participant will have a fairly good idea of what his annual pension benefits will be when he retires. Second, in most cases, an annual pension benefit is also indexed for inflation, also discussed in Chapter 2. Therefore, he does not have to worry about losing the purchasing power of his pension benefit to inflation in the many years he will continue to live after retirement. Third, the retiree in a DB plan does not have to be worried about the financial implication of longevity. He is guaranteed pension benefits for as long as he lives. In this sense, a DB plan serves more like a group insurance program on longevity. When a DB plan has enough members, it can use average mortality for long-term financial planning purposes. In this way, a member who lives longer than an average life span does not have to worry that the plan will run out of financial resources to fund his longer life span, as it will be balanced out by other members with life spans shorter than average. In other words, each employee has a guaranteed lifetime annuity.

A second advantage of a DB plan accrues to the employer/plan sponsor. A DB plan rewards longevity of career with a single employer because, again discussed in Chapter 2, pension benefits accrue faster the longer an employee stays with one employer. Although not necessarily its intended purpose, a DB plan helps retain current employees.

The third main advantage of a DB plan is the cost of management. Because of the size of most DB plans in the public sector, they can benefit from economies of scale in lowering management costs. As pension assets are invested for the long haul, a relatively small decrease in management cost can lead to significant savings over time. Such savings can be realized in the form of either lower pension contribution from the employer or employee, or higher pension benefit to the beneficiary. Such cost savings, however, can be offset to some extent by the expenses related to the administration of pension benefits, such as actuarial valuation and investment consulting.

7.1.2 Disadvantages of a Defined Benefit Plan

The biggest disadvantage of a DB plan is the funding risk for the plan's government sponsor. The funding risk refers to both short-term volatility in pension contribution and long-term uncertainty in required funding to meet future pension obligation. This funding risk results from several assumptions built into the calculation of costs related to funding pension benefits. The first and most important assumption is the assumed rate of return on the investment. As discussed in Chapter 3, this is the rate used to determine the present value of pension liability, which is then compared to the value of pension assets to arrive at the funding ratio of the plan. Plan assets are expected to earn the assumed rate of return in order for the plan to be fully funded and for the annual contribution to remain steady. There are two risks involved in the rate of return assumption. First, since the assumption is made for a long-term average rate of return, the yearly return will most likely be different from this average return, which will lead to ups and downs in the value of pension assets. This will cause the pension plan to be either overfunded when the actual return is substantially higher than the assumed return, or underfunded when returns are lower than the assumed rate. Without compensating mechanisms, these fluctuations would cause the pension contribution to go up and down, leading to budgetary volatility for the plan sponsor. While smoothing techniques in valuing pension assets can reduce some of the volatility in asset value caused by the investment returns, they will not completely eliminate it in extreme cases. The best examples of this volatility, despite smoothing techniques, are the changes in funding ratio and pension contribution witnessed since the mid-1990s, as shown in Table 1.4 and Figure 1.4 in Chapter 1.

The second risk with regard to the rate of return is whether it can be realized in the long run. Because the assumed rate of return is long-term in nature and is based on historical rates of return, there is always the potential that this assumed rate of return might not be realized in the future. If the plan's actuaries decide that the assumed rate of return cannot be achieved over time and has to lower the assumed return, then the plan sponsor's contribution has to go up to make up for the difference.

Another set of assumptions that can cause uncertainty in pension funding is related to demographics. Pension benefits are calculated using many demographic assumptions, such as when people retire and how long people will live after they retire. If any of these assumptions turn out to be different from the real world experience, the pension contribution level will need to be adjusted. For example, if retirees as a group live longer in the future than assumed, then the pension liability will have to go up. This is similar to the situation facing the nation's Social Security program.

While the funding risk due to these actuarial assumptions is somewhat beyond the plan sponsor's control, there is another kind of funding risk that can be caused by the plan sponsor itself. This has to do with the political nature of government finance. As discussed in Chapter 6, pension benefit is essentially part of the cost of providing public services. This cost is different from other costs, however, in that while the benefit to the employee is measurable, the full cost of providing the pension benefits to the plan sponsor is spread out over many years and, thus, the immediate cost is typically very small. This creates a potential for a moral hazard problem. Moral hazard means that an agent taking an action is not responsible for the consequences of the action, which may be detrimental to the interest of the principals who hire the agents. In the case of public pension funding, the principals are taxpayers who are ultimately responsible for paying the public pension benefits. It is tempting for current elected officials who serve as the taxpayers' agents, to promise pension benefit increases to public employees without actually paying for the full cost upfront. By promising increases, the current elected officials will take the credit for improved pensions without having to determine upfront how to pay for them. It will be up to future elected officials and taxpayers to pay for the full cost. This adds financial burden to the future generation and creates funding risk.

Moral hazard problems also occur when the current elected officials do not fully fund the required annual pension contribution. Unlike other government-funded programs, failure to appropriate full actuarial funding does not lead to immediate cuts in service; even if the pension contribution is not fully funded in that year, there should be sufficient assets in the plan to cover any immediate benefit payments. Therefore, there is also temptation for a plan sponsor to forego pension contributions when it faces budgetary stress. The sponsor is pressed to fund other more important government programs and a pension fund can be a convenient place to extract some extra funds. The consequence of such a fiscal gimmick is that the pension cost has been pushed into the future. If this is a one-time action, there are minimal long-term financial consequences. If the government fails to fund over a long period of time, then it will create a severe financial burden for future taxpayers.

In addition to plan sponsors, the actions taken by plan members can also be a source of funding risk, due to the formula used to calculate final pension benefits.

Since pension benefits are based on the salaries of the final few years, the employee can take some action, such as earning more overtime, to create a spike in salaries in these final few years to significantly boost his pension benefits. Since all of his future benefits, including cost-of-living adjustments, will be based on the initial pension benefit amount, a small increase in this initial amount can lead to significant cost increases over time.

While funding risks are the major disadvantages of a DB plan, another disadvantage of a DB plan is the lack of portability for the plan participants, as a result of vesting requirements and backloading of pension benefit accruals. However, many features of public pension benefits have reduced this disadvantage to public plan participants:

- Purchase of service credits
- Availability of a state-wide pension plan in many states that is open to all state and local government employers, which allows employees to change jobs within the same state without any loss in pension benefit accrual
- Reciprocal agreement among different pension systems, primarily within the same state. In such an agreement, when an employee changes job and switches from one retirement plan to another, his service credit will also be transferred from his previous plan to his current retirement plan. For example, Illinois has a law known as the Reciprocal Act that covers 13 public retirement systems in the state. Each system "recognizes" service earned in another system for purposes of the participant's vesting and final average salary calculation.

7.1.3 Some Evidence of Defined Benefit Plan Risks

State and local public pension plans have been on a roller coast ride since the early 1990s. Table 1.5 in Chapter 1 shows the level of volatility in total state and local government pension contributions since the late 1990s. Between 1996 and 2002, total annual government contribution was reduced by 13 percent. In the next two years, plan sponsors were forced to increase contributions by more than 50 percent just when their portfolios were feeling the full brunt of the stock market decline.

There is also evidence to suggest that there is some linkage between pension funding status and increases in pension benefits. Benefit increases come in many ways. The most common way is to increase the multiplying factor in the pension benefit formula. Based on Wisconsin's survey of 83 major state-level pension plans, the number of pension systems increasing the multiplier increased with the improvement in funding ratio. For example, between 1994 and 1996, when the average funding ratio was still below 90 percent, only nine pension systems increased the multiplying factor. Over the next four years, when the average funding ratio for public pension plans kept improving, the multiplying factor was increased in 27 plans. In 2000, the average funding ratio of public plans reached its peak of 100 percent, with many pension plans' funding ratio over 100 percent. Not surprisingly, over the two-year period between 2000 and 2002, 21 plans increased their multiplying factor. When the stock market declined, which reduced the funding ratio of most pension plans, in some cases by a significant margin, benefit increases slowed down and only six plans increased their multiplying factors between 2002 and 2004. Overall, 41 out of 83 plans increased their multiplying factors over the 10-year period between 1994 and 2004, with some of them increasing the factor several times during this period. Since these benefit increases are almost always retroactive, such increases lead to a substantial jump in future pension liabilities.

Other ways of increasing pension benefits include shortening the period in determining the final average salary (FAS) and reducing the age or service requirement for normal retirement benefits. Shortening the FAS period increases the salary base for determining the pension benefits. For example, state governments in Kentucky and Missouri shortened the period from five to three years; and California reduced the averaging period from three years to just the final year for members of its two giant pension systems. Reducing the age and service requirement for normal retirement benefits both reduces the time over which more pension assets can grow and increases the time over which pension benefits will be withdrawn. For example, the California state government reduced the normal retirement age from 60 to 55 with five years of service, the Georgia state government reduced the age from 65 to 60, and Massachusetts reduced the age from 65 to 55.

When pension contribution deduction is combined with pension benefit increase in multiple forms, the effect on pension funding ratio and subsequent pension contribution increase can be overwhelming, as in the case of California. In 1999, the funding ratio of California Public Employees Retirement System (CalPERS) (2006) reached a peak of 128 percent. Along with the pension benefit increases mentioned earlier, pension contribution was also reduced from $2.3 billion in 1998 to $321 million in 2001. These actions and the stock market decline reduced the CalPERS's funding ratio to 87 percent in 2005, a decline of 40 percentage points from its peak. Accordingly, pension contribution was also increased to $6.1 billion in 2006.

7.2 Defined Contribution (DC)

The defined contribution (DC) plan has become a widely used plan type in the private sector after the establishment of the 401(k) individual retirement plan in 1981, as discussed in Chapter 1. At the same time, the membership in the defined benefit plan type has been on the decline. In the public sector, people working in higher education institutions traditionally participate in DC plans.

7.2.1 Advantages of the Defined Contribution Plan

The DC plan eliminates some of the disadvantages of a DB plan. First and foremost, the DC plan addresses the funding risks to the plan sponsor of the DB plan. As discussed in the previous section, such funding risks result from the economic and demographic assumptions in the actuarial valuation as well as actions taken by both plan sponsors and members. All of these sources of funding risk are inoperative in a DC plan. First, the economic and demographic assumptions are no longer relevant to plan sponsors as they do not guarantee pension benefits and, therefore, have no need to engage in actuarial valuation. Second, employers can no longer promise pension benefit increases without considering the cost. In a DC plan, the only way to increase pension benefits is for the sponsor to increase the contribution to a member's retirement account. Therefore, the cost is immediate, easily observable, and evenly spread out over the service years of the plan member. Plan sponsors also no longer have the luxury of delaying pension contributions to members' retirement accounts, as they can in a DB plan. Third, there is no longer any incentive for plan members to engage in actions that will increase salary in the final years, as such action will not have any significant effect on their pension benefits. In short, the biggest advantage of a DC plan is the funding certainty to the plan sponsor. As long as the plan sponsor fully funds pension contributions to the members' retirement accounts, which are set at a constant percentage of salary from year to year, they will no longer face any funding risk and unfunded pension liability in the future. This creates short-term and long-term budgetary certainty for the plan sponsor.

While eliminating funding risk is the major advantage of a DC plan, another advantage is its portability. The DC plan is more portable than the DB plan because all benefits are accumulated in the member's individual retirement account. More importantly, unlike the DB plan in which the bulk of the pension benefits are accrued toward the end of the career, the accrual of pension benefits is more evenly spread out over one's career in a DC plan. The accrual of pension benefits will no longer be affected by whether an employee stays with one employer for his entire career or switches jobs multiple times in his career. When he terminates one employment, he can simply take the individual retirement account to his next employment. The potential disadvantage of such portability is that it makes it much easier for an employee who terminates prior to retirement to take a lump sum distribution from his retirement account and use the money for purposes other than savings toward retirement.

7.2.2 Disadvantages of the Defined Contribution Plan

These two major advantages to the plan sponsor and members also bring potential disadvantages to plan members.

7.2.2.1 Pension Benefit Certainty

The major disadvantage is the uncertainty of pension benefits to plan members. There are three major sources for this uncertainty. First and foremost is the investment return on the individual retirement account, which is entirely the responsibility of the plan member. As pension assets are invested for many decades, the member's retirement benefits are far more dependent on investment income than on the periodic contribution from the employer and employee. If the employee is unable to achieve an adequate long-term rate of return, then she runs the risk of not having enough assets in this account to fund her retirement benefits. In particular, poor investment returns during the first few years of retirement can wreck havoc with the retiree's accumulated assets and her financial security in retirement. In this case, her pension benefits may not bear any relationship to preretirement salary, as in the case of a defined benefit plan.

The second source of uncertainty of retirement income is life expectancy. Average life expectancy at age 65 is 16 years for men and 19 years for women (Board of Trustees, 2004). Because this is an average life expectancy, half of all retirees will live many years beyond the average life expectancy. Therefore, it is more difficult to gauge what level of financial resources a retiree needs to fund his or her retirement. The defined contribution plan member bears the financial risk of outliving accumulated assets.

One way to eliminate this source of uncertainty is for the member to purchase an annuity from a financial company at the time of retirement with the assets in the individual account. There are several common annuity options, including a single or joint annuity, either of which can be a life-time or period certain annuity:

1. A single lifetime annuity pays the retiree a fixed payment until he dies. The retiree's heirs do not receive anything else. The risk is that if the retiree lives a short time after the retirement, then he loses all the money to the insurance company. Despite such risk, single life annuities are still a popular choice for people who want to be absolutely sure they don't outlive their assets.
2. A joint and survivor annuity is based on two lives. The payments continue until both people die. With a joint-life annuity, the retiree can specify the percentage, such as 50 or 75 percent of the initial annuity payment that the survivor will receive after the first spouse dies. The payment in such an annuity is lower than that in a single life annuity.
3. Period certain annuities guarantee payments to an annuitant for a particular period of time. For example, a 10-year period-certain annuity will make annuity payments for 10 years and no more. If the annuitant dies before the 10 years have expired, the payments will continue to the policy's beneficiaries for the remaining years. The monthly payment rate for a period-certain annuity is generally higher than that for a life annuity.

Single life annuities with a period certain are a combination of a life annuity and a period certain annuity. For example, if a retiree purchases a life annuity with a 10-year period certain, he will receive payments for the rest of his life — no matter how long he lives. However, if he dies shortly after he starts receiving the annuity, the policy's beneficiaries will receive payments until 10 years have expired.

While the life annuity and its variations can limit the risk that the retiree faces due to life span, there are, however, two uncertainties related to this approach. As suggested earlier, the value of annuitized income that a retiree can purchase will be dependent on the amount of assets in his or her individual retirement account and the prevailing interest rate. First, even if the employee manages to earn a decent rate of return while she is working, a financial market downturn at the time of or shortly before her retirement can significantly reduce the amount of assets that she will use to purchase a lifetime annuity. Second, as the purchase price is equal to the present value of this lifetime stream of annuity payments, with the discount rate being closely linked to interest rates at the time of purchase, the size of the annual annuity will also depend on the interest rate environment at the time of retirement. If interest rates are relatively low, then the retiree will receive a smaller annuity, everything else being equal.

It is worthwhile to compare the DC annuity concept to the defined benefit pension benefit to gain a better understanding of the differences between these two pension plans. In essence, the pension benefit to a retiree in a defined benefit plan is also in the form of a lifetime annuity, except that the retiree does not have to worry about the investment performance and market interest rate at the time of retirement. The advantages enjoyed by a defined benefit plan allow it to ignore the effect of short-term investment performance and interest rate volatility when offering this lifetime annuity to retirees. Due to the group insurance nature of a large public pension plan, short-term investment performance and interest rate do not play a role in determining the size of the annuity for individual retirees. The investment horizon of a defined benefit plan is much longer than that of an individual retiree in a defined contribution plan. As a matter of fact, the investment horizon of a defined benefit plan in the public sector is effectively infinite, since most plans are expected to exist in perpetuity. Given this infinite time horizon, the defined benefit plan can afford to ignore poor investment return or decreases in pension asset value in the short run because it can recover from such losses with sufficient time. Moreover, not everyone is retiring at the same time and new pension contributions continue to come into the defined benefit plan. Therefore, a defined benefit plan can ignore the effect of short-term market performance when offering retirees lifetime annuities.

Another advantage enjoyed by a defined benefit plan in offering a lifetime annuity is the cost. The retiree in a defined contribution plan will have to purchase an annuity from a financial company whose motive in selling this product is presumably to make a profit. The annuity in a defined benefit plan is offered by the plan sponsor itself with no apparent incentive to make any profit from the retiree. Therefore, everything else being equal, it is most likely that for the same lifetime annuity, a defined benefit plan can offer it at a lower cost to a retiree than a financial company can to a retiree in a defined contribution plan. This cost saving can be turned into a larger annuity for the retiree or reduced contributions for the sponsor in a defined benefit plan.

The third source of uncertainty for the retiree in a defined contribution plan is future inflation. As inflation reduces the purchasing power of any financial assets, a retiree in a defined contribution plan needs to make sure that he has sufficient assets and that any remaining assets will keep growing in value along with inflation, which adds another layer of complexity to his financial planning. It is possible for the retiree to purchase an annuity indexed for inflation, but at a substantially higher cost. As discussed in Chapter 2, most retirees in defined benefit plans are not concerned about inflation because their benefits are adjusted for inflation each year.

7.2.2.2 Investment

While the uncertainty (or more appropriately, the sustainability) of retirement income is the most important concern for a member in a defined contribution plan, there are also potential disadvantages with regard to investment in the areas of investment management cost, investment knowledge and discipline, and diversification:

1. An individual investment account, due to its small size, will most likely incur a higher management cost than a large DB plan, which can benefit from the economy of scale. As mentioned earlier, a small difference in management fee as a percentage of assets over time can lead to a significant difference in the value of accumulated assets.
2. A successful long-term investment requires both investment knowledge and discipline. Investment knowledge is required to design a proper long-term asset allocation strategy that is neither too conservative nor too risky. Investment discipline is required so the investor can stick to the long-term investment strategy through the ups and downs of the financial market. If he lacks either the knowledge, or discipline, or both, then it increases the chance that he will not be able to achieve his investment goals.
3. Even if the DC plan member has the knowledge and discipline, his investment portfolio will most likely not be as diversified as that of a large public pension plan, as some investment options are not available to individual investors because of the relatively small size of the individual account. For example, only large public pension funds are able to purchase alternative investment instruments, such as private equity funds, hedge funds, and direct investment in real estate properties. As discussed in Chapter 5, because these investment instruments tend to have low correlation with traditional investment options,

such as stocks and bonds, a small allocation to these assets can provide further diversification for the entire portfolio.

Due to the various potential drawbacks of the defined contribution plan, many research studies have shown that private sector DC participants on average have not achieved investment returns equal to those achieved by large pension funds. According to a study conducted by Boston College between 1988 and 2004, the median rate of returns for defined benefit and 401(k) plans were 10.7 percent and 9.7 percent, respectively (Munnell, et al., 2006). The authors attributed this difference of 1 percent to higher investment cost, poor market timing, and investment mistakes. The authors also found that nearly half of 401(k) participants lacked diversification, either fully invested in stocks or not invested in stock at all.

7.2.2.3 Disability Benefit

A final disadvantage of the DC plan is the lack of specific provisions for disability benefits. Most defined benefit plans pay disability benefits to individuals who become permanently disabled and cannot work. The cost of providing such benefits is already included in plan funding. Most DC plans, however, do not provide for such benefits. If the DC plan does not have a separate disability insurance policy program, then a disabled DC plan member can only have the balance in her retirement account to draw on. This disadvantage is at its worst when the defined contribution plan member is disabled early on in her career because the balance in the account can be fairly modest. As discussed in Chapter 2, defined benefit plans set a minimum disability benefit, expressed as a percentage of an employee's predisability salary, to guarantee a minimum living standard. Thus, the individual retirement account could provide significantly less income than a disability benefit under a defined benefit plan. This difference, however, will gradually disappear if the defined contribution plan member becomes disabled later in her career. The employee can also purchase long-term disability benefits on her own from an insurance company. The downside is the higher cost compared to a defined benefit plan.

7.2.3 Summary

To summarize, the biggest disadvantage of a defined benefit plan for the plan sponsor is its funding uncertainty, both short-term and long-term. The biggest disadvantage of a defined contribution plan for the plan member is the uncertainty of retirement income, both at the time of retirement and after retirement. This can be viewed as the funding risk from a DC plan member's perspective. These two funding risks are closely related to each other. By switching from a defined benefit plan to a defined contribution plan, the plan sponsor essentially shifts the funding risk to the plan members.

7.3 Public Sector Defined Contribution Plan Experience

Due to the various risks associated with defined contribution plans as discussed earlier, most of the calls for switching to DC plans have run into significant resistance from public employees, from unions who represent public employees, and from some elected officials who view the shift to a DC plan as a threat to retirement security. As mentioned in Chapter 1, about 90 percent of employees in the public sector are in defined benefit plans and this number has changed very little over the years. This is quite different from the private sector, which has seen a steady increase in the percentage of employees participating in defined contribution plans. The low defined contribution participation rate in the public sector, however, does not mean that only a small percentage of people in the public sector have the option to participate in DC plans. As a matter of fact, many states offer optional DC plans. What differentiates the public sector DC plans from those of the private sector is that in the public sector, the DC plan is mandated for only a very small number of people.

Various DC plans in the public sector can be divided into four groups based on two features of the plan: (1) whether the participation is mandatory or optional and (2) whether the plan is an exclusive DC plan or a hybrid plan.*

7.3.1 Mandatory Participation in Exclusive DC Plan

This is the most stringent of the four groups. In this group, participation in a defined contribution plan is mandatory and it is the only plan option available to the member. As of 2006, there are only two major state-level pension systems in this group.

In Michigan, all employees hired after 1997 must participate in a defined contribution plan. The state of Michigan contributes an amount equal to 4 percent of an employee's gross salary to the plan participant's individual account. In addition, the state will also match any contributions the employee makes to his account, up to another 3 percent per pay period (State of Michigan, 2004). There is a vesting requirement for the state contribution: 50 percent vested after two years, 75 percent vested after three years, and fully vested after four years. The Michigan DC plan also provides disability benefits. If an employee becomes permanently disabled due to an injury or illness incurred at work, the state will pay a disability benefit regardless of how long he has been employed by the state. If the injury or illness is not incurred at work, disability benefits may be payable only if the employee has the equivalent of 10 years of state employment.

---

* A hybrid plan in this context means a plan has two parts in it, a defined benefit part and a defined contribution part.

The other mandatory and exclusive defined contribution plan is in Alaska. In 2005, the Alaska state legislature passed legislation that required all employees hired after July 1, 2006 to participate only in a defined contribution plan. The employee is required to contribute 8 percent of his salary and the state will contribute 5 percent. There is also a vesting requirement for state contribution: 25 percent vesting after two years, 50 percent after three years, 75 percent after four years, and fully vested after five years. The defined contribution plan also pays occupational disability benefits, equal to 40 percent of the salary immediately before the termination of employment due to the disability. It also provides occupational death benefits to survivors, equal to 50 percent of salary prior to death of police and firefighters and 40 percent for all other occupations.

Two other mandatory and exclusive defined contribution plans were discontinued in recent years. Nebraska used to have a stand-alone mandatory DC plan for state and county general employees. It was discontinued at the end of 2002 and was replaced by a cash balance plan.* In West Virginia, all school employees hired after January 1, 1991 were automatically enrolled in that state's exclusive DC plan for teachers. Due to the power investment performance of DC member, the DC plan was closed to new participants after June 30, 2005 and the traditional defined benefit plan again became the retirement plan for employees hired after that date.

7.3.2 Optional Participation in Exclusive Defined Contribution Plan

In this group, the public employee has the option to participate exclusively in a defined contribution plan or stay in a traditional defined benefit plan. Following is a brief description of DC plans in this group:

- **Colorado**: In 2004, Colorado created a defined contribution plan as an option for state employees, effective January 1, 2006. The employee contributes 8 percent and the employer contributes 10.15 percent. Employer contributions become fully vested in the plan after completion of five years of participation according to the following schedule: 50 percent immediately, 60 percent after one year, 70 percent after two years, 80 percent after three years, 90 percent after four years, and fully vested after five years.
- **Florida**: Florida created an optional defined contribution plan in 2002, called the Florida Retirement System Investment Plan. All employees, old and new, can opt to participate in the DC plan. This option is irrevocable after a certain period of participation. Employer contribution rates range from 9 to 20 percent of salary depending on membership class. Vesting in employer contribution occurs after one year.

* Cash balance plan will be discussed in detail in the following section.

- **Montana**: Montana's Defined Contribution Retirement Program (DCRP) is a multiple employer plan established on July 1, 2002. This plan covers eligible employees of the state, university system, local government, and certain employees of the school districts that elect the defined contribution plan. All new employees are members of the Defined Benefit Retirement Program (DBRP) initially. They have a 12-month window during which they may choose to transfer to the DCRP or remain in the DBRP. The choice is irrevocable. The employee contributes 6.9 percent of his salary to his individual account and the employer contribution rate is 4.19 percent of employee salary. The vesting period for employer contributions is five years. In addition, the employer also contributes 0.30 percent of employee salary to a long-term disability fund. An employee can receive disability benefits in two ways, either receiving it from the Long-term Disability Trust Fund or drawing from the balance in his individual account, but not both.
- **Ohio**: Effective July 2001, new employees and nonvested members as of that date in the Ohio Public Employees' Retirement System and State Teachers' Retirement System of Ohio can opt to participate exclusively in the respective system's defined contribution plan. The employee contribution rate is 9 percent of earnable salary. The employer contribution rate is 13.54 percent of earnable salary for the state division and 13.7 percent for the local division. The vesting schedule for employer contributions is as follows: 20 percent after one year, 40 percent after two years, 60 percent after three years, 80 percent after four years, and full vesting after five years. The Ohio defined contribution plans do not provide disability benefit.
- **South Carolina**: In South Carolina, all state, public school, and higher education employees hired after June 30, 2001 can elect to participate exclusively in a defined contribution plan called South Carolina State Optional Retirement Plan. Prior to July 1, 2006, the employer contribution rate was 8.05 percent of employee earnings and after July 1, 2007, the employer contribution rate is 8.55 percent of employee earnings. There is no vesting period. The South Carolina DC plan does not provide disability benefit.

7.3.3 Mandatory Participation in Hybrid Plan

In this group, employees must participate in a hybrid plan that consists of an employer-funded defined benefit component and an employee-funded defined contribution component. Because the hybrid plan contains a defined benefit component, disability benefits are available to plan members.

Indiana Public Employees' Retirement Fund and State Teachers' Retirement Fund are two examples. They have two components: the DB pension plan and the DC Annuity Savings Account. The DB plan is entirely funded by the employer. As for the DC part, the state law requires that 3 percent of an employee's gross wages (regular and overtime pay) must be contributed to fund the Annuity Savings Account. The employee may also be able to make additional contributions to his Annuity Savings Account if the employer has allowed for that (Indiana Public Employees' Retirement Fund).

Oregon is the other state that mandates participation in a hybrid plan. The 72nd Oregon Legislature created the Oregon Public Service Retirement Plan (Oregon Public Employees Retirement System). Public employees hired on or after August 29, 2003 become part of OPSRP. OPSRP is a hybrid pension plan with two components: the Pension Program (the defined benefit part) and the Individual Account Program (IAP) (the defined contribution part). Employer's contributions go to fund the DB program. OPSRP members contribute 6 percent of their salary to the IAP, and the specific government agency, which hires the employee, may agree to pay the 6 percent contribution.

7.3.4 Optional Participation in the Hybrid Plan

In this group, the state gives the active member a choice to join either the traditional defined benefit plan or a hybrid plan.

In addition to the optional defined contribution plan, Ohio also has an optional hybrid plan, call the Combined Plan. Under the defined benefit portion of the Combined Plan, which is funded by the employer contribution, the member's retirement benefit is determined by a reduced formula. The percentage-multiplying factor is 1 percent in the Combined Plan formula, whereas it is 2.2 percent in Ohio's traditional defined benefit Plan (Ohio Public Employees Retirement System, 2007). Under the defined contribution portion of the Combined Plan, employee contributions are deposited into the member's individual account.

In 2000, the Washington State Legislature created the Public Employees' Retirement System (PERS) Plan 3, a hybrid plan modeled after Washington State's Teachers' (TRS) and School Employees' Retirement System (SERS) Plan 3 created in 1996 and 2000 respectively (State of Washington). Within PERS Plan 3, member contributions go into defined contribution individual accounts and employer contributions go to fund the defined benefit component. New PERS employees must choose to become either Plan 2 (a traditional DB plan) or Plan 3 members within 90 days of hiring. If the employee does not select a plan within 90 days, the default plan becomes Plan 3. Plan choice is irrevocable. Established PERS Plan 2 members (those who first entered service prior to the implementation of PERS Plan 3) can choose to transfer to Plan 3 each year during the month of January. Such transfer to Plan 3 is also irrevocable.

	Mandatory	Optional
Exclusive DC Plan	Michigan Alaska	Colorado, Florida, Ohio, Montana, South Carolina
Hybrid Plan	Indiana, Oregon	Ohio, Washington

Figure 7.1 Public-defined contribution and hybrid plans.

7.3.5 Summary

Figure 7.1 provides a summary of these various plans described above. Based on this brief description of various defined contribution plans, it is easy to see that, unlike the private sector, mandatory participation in exclusive DC plans is very rare in the public sector. The most telling sign of the popularity of a DC plan can be found in the number of employees who choose the DC plan in the optional participation in exclusive or hybrid plan groups. In all of these plans, well over 90 percent of new employees and employees already in the traditional plan, who were given the option to switch to a newly created defined contribution plan, decided to stay with the traditional defined benefit plan (Sostek, 2004; Gosselin, 2005).

Why did so few public employees choose'to participate in a DC plan when given the choice? One important reason has to do with the investment environment. Many of the DC plans were created after the stock market decline from 2000 to 2002. Even though the stock market came back strongly in 2003 and 2004, many would-be DC participants were deterred by such risks and volatility in investments and, thus, wanted to seek more security in the traditional DB plan. Had these plans been created in the late 1990s, it is possible more people would have made the switch to the DC plan. This fear of investment risk can also be shown in the asset allocation strategy of those employees who have to participate in a DC plan. A major reason that the mandatory DC plan for teachers was discontinued in West Virginia was that the teachers invested very conservatively with a significant portion in money market funds, leading to very low returns. The reason for the discontinuation of the mandatory DC plan in Nebraska has also to do with the poor performance of plan members' investments. Such poor investment performance is by no means just limited to DC plan members in the public sector.

7.4 Cash Balance Plan

Because the switch from defined benefit to defined contribution plan essentially shifts the funding risk from plan sponsors to plan members, if financial well-being is considered in an aggregate for both plan sponsors and plan members, then one plan holds no distinctive advantage over the other when it comes to reducing funding risk. There is, however, an alternative pension plan, the cash balance (CB) plan, that allows the plan sponsors and plan members to share the funding risk. The cash balance plan is primarily a defined benefit plan as the benefits at the time of retirement can be roughly determined and guaranteed by the CB plan sponsor. However, a CB plan also takes on some of the characteristics of a defined contribution plan. Therefore, a CB plan is also called a hybrid plan. The CB plan has been in existence and widely used in the private sector for many years, whereas its usage in the public sector has been very limited. In this section, we discuss what a CB plan is and how it addresses the funding risks facing both defined benefit and defined contribution plans. CB plans in Nebraska and California are used to illustrate how they work in the public sector.

7.4.1 Cash Balance Plan

There are four major plan features that define a CB plan.

First, in a CB plan, a beneficiary's final pension benefits when he retires are based on the employer's and employee's pension contributions, plus a guaranteed minimum rate of return on the investment. Even though the benefit in a CB plan is still the defined benefit type, the determination of pension benefit level is quite different from that in a defined benefit plan. In a DB plan, the benefit level is based on the number of years of service, the final average salary, and a multiplying factor, regardless of how much the employer and employee have contributed in the past and how the investment has performed over the employee's career. It is also different from the defined contribution benefit determination. In a DC plan, a member's benefit is determined by the employer's and the member's contributions in the past, plus the investment return on the contributions, which varies from one member to the other depending on how they invest. The main difference between CB and DC plans in terms of benefit level determination is that the investment returns on pension contributions are guaranteed by plan sponsors of CB plans, but not guaranteed by plan sponsors of DC plans.

Second, each member maintains a notional individual account in the CB plan that tells the member the exact amount of pension benefits accrued at any given time up to the time he retires. This is a defined contribution plan characteristic. It is a notional account because in the actual CB plan, there is no separate account with the individual member's name on it. In this notional account, the balance increases by the annual employer and member contributions, plus a guaranteed minimum annual interest rate credit on the previous year's account balance. Cash balance plans generally credit interest to participant hypothetical accounts using an index tied to a Treasury security (Governmental Accountability Office, 2000). If the actual investment return is higher than the minimum guaranteed return on the notional account, the CB plan will also give the plan member a one-time extra interest credit that will increase his account balance. Even though the granting of extra interest credit is just a one-time event, its impact is permanent, as all future interest credit will be based on a larger account balance.

Third, unlike the defined contribution plan in which the individual members manage their own investment within their retirement accounts, members of CB plans do not manage the investment of assets in their notional accounts. All assets in those notional accounts are invested and managed by the administrator of CB plan, just like in a defined benefit plan.

Fourth, when a member terminates his employment, he can take the balance in his notional account with him to his new employment, just like in a defined contribution plan. When he retires, he has two options. He can take out the entire balance in his notional account in one lump sum and manage the assets on his own from then on, or he can take various kinds of annuities offered by the plan sponsor, with the present value of the annuity equal to the balance in his notional account.

With these four basic characteristics of the CB plan, we can proceed to analyze how it addresses the disadvantages of both DB and DC plans.

7.4.2 Cash Balance Plan versus Defined Benefit Plan

The cash balance plan addresses many of the disadvantages of defined benefit plans to both the plan sponsors and plan members:

1. The guaranteed rate of return on the members' notional accounts is much lower than the assumed rate of return for an average defined benefit plan, a factor that reduces the funding risk for plan sponsors. The typical guaranteed rate of return for most CB plans, tied to the Treasury securities yield, is 5 percent. This is significantly lower than the assumed rate of return of about 8 percent for most public defined benefit plans. The long-term risk of not being able to achieve the guaranteed rate for a CB plan, therefore, is a lot smaller than for a DB plan. It is still possible that even this lower guaranteed rate of return may not be achieved for a significant period of time. However, with a surplus reserve fund set aside that holds the part of the surplus return over the guaranteed return in the past, the negative impact of such return should be fairly limited.
2. As the retiree's pension benefit is based on the balance of his notional account rather than a formula, there is no long-term funding risk to the plan sponsor due to changes in demographic assumptions, such as increase in longevity, as long as the contributions are made and the minimum guaranteed investment return is achieved. It also makes it less likely for a "spike" in final salary to occur, as it will have almost no impact on the final pension benefit for the plan member. This also removes another source of potential funding risk for the plan sponsor.

3. The CB plan is much less susceptible to moral hazard problems, as the pension contribution is a guaranteed percentage of the member's salary and the benefits are based on the final account balance. If elected officials want to increase pension benefits, they have to increase the employer's contribution rate. The increase in pension cost of such a measure to the government, however, is immediate and evenly spread out in the future, making it less appealing to elected officials.
4. Cash balance plans provide more uniform accrual of benefits throughout the employee's career, when compared to the accrual of benefits in a defined benefit plan. Job change, thus, has no negative consequence on pension benefit accrual for a CB plan member. This also solves portability issues of defined benefit plans due to frequent job changes.

7.4.3 Cash Balance Plan versus Defined Contribution Plan

The cash balance plan also eliminates the sources of funding risk to defined contribution plan members discussed earlier:

1. In cash balance plans, the investment decisions and the investment performance are the responsibility of the employer, not the employee. Even though the benefits are based on individual accounts, assets are managed in aggregate by the plan trustee. Thus, employees are unable to make any grave investment errors that can cause their long-term investment performance to suffer. Investment cost should also be lower for a cash balance plan.
2. Cash balance plans must offer employees the ability, within the plan, to convert their account balances to lifetime annuities at no additional cost. This also eliminates another uncertainty related to annuity for a defined contribution annuitant. It is possible for the cash balance plan to offer the annuity as the assets are centrally managed by the sponsor, whereas in a DC plan, the assets are managed individually. One major difference between the lifetime annuity for a CB plan member versus that for a DC plan member is that the final account balance of a CB plan prior to its conversion to an annuity is not affected by the financial market at the time of conversion. As discussed earlier, one of the disadvantages of a DC plan is the uncertainty of account balance at the time of conversion due to the financial market fluctuation. Because the CB plan guarantees a rate of return, the member does not have to worry how the fluctuating financial market affects the balance. The CB plan is able to make such a guarantee regardless of what the market return is because of the group insurance characteristics of the CB plan. Also, because the annuity is offered by the CB plan, it should be less expensive than the same annuity purchased from a commercial company.

7.4.4 Summary

Table 7.1 compares the three plan types in terms of the major features that define a pension plan.

The central question becomes: How can a cash balance plan address the funding risks to both the plan sponsors of defined benefit and members of defined contribution plans? The answer to this question is that the funding risk does not simply disappear in a cash balance plan, but it is now shared between the plan sponsor and member. By guaranteeing an investment return of around 5 percent to plan members rather than 8 percent, the plan sponsor significantly reduces its own funding risk if the higher investment return cannot be realized. While the guaranteed minimum retirement benefits in a CB plan are most likely to be lower than those in a DB plan due to the lower guaranteed investment return, there is also a chance for increased benefits if the actual return turns out to be higher than

Table 7.1 Comparison between DB, DC, and CB Plans

	Defined Benefit	*Defined Contribution*	*Cash Balance*
Pension benefit	Based on a formula, determined by years of service, final salary and multiplying factor	Based on individual investment account balance, determined by past contribution and investment return	Based on a notional individual account balance, determined by contribution and a guaranteed minimum investment return
Investment	Plan sponsor is responsible	Plan participant is responsible	Plan sponsor is responsible
Funding risk to the plan sponsor	Assumes all funding risk due to uncertainty in investment and future funding requirement	Assumes no funding risk	Assumes limited funding risk
Funding risk to the employee	Assumes no funding risk	Assumes all funding risk due to uncertainty in investment and amount of assets required in retirement	Assumes no funding risk
Moral hazard	Potential for moral hazard	No potential for moral hazard	Limited potential for moral hazard.

the guaranteed minimum return. In this case, the plan sponsor will also share the additional investment returns with CB plan members by increasing the interest rate credit in the years when there is surplus return. In other words, beyond a minimum guarantee, retirement benefits in a CB plan are tied to actual investment performance, which lies at the heart of the risk sharing between plan sponsors and plan members.

Due to the various advantages of CB plans, many companies in the private sector replaced their traditional DB plans with CB plans in the 1980s and 1990s. Its further advancement, however, was hampered by legal issues. When a DB plan is changed into a CB plan, older workers will be more negatively affected by the change than younger workers in terms of pension benefit accrual, largely because the pension benefit accrual accelerates in the later years of one's career in a DB plan, but is more evenly spread out over one's career in a CB plan. Lawsuits were filed against CB plans on the grounds of age discrimination. Companies that would have switched to CB plans stopped doing so due to the legality of these plans. Such legal issues, however, never affect public pension plans. As discussed in Chapter 4, public pension rights of current employees are protected and any new plan or benefit changes only apply to new employees. Despite the lack of any legal issues surrounding CB plans in the public sector, its use remains extremely limited. So far, the cash balance option is being used by only one large state pension plan (Nebraska) and also being used for a very limited group of employees in California.

7.4.5 Nebraska Cash Balance Plan

The establishment of the cash balance plan in Nebraska offers an interesting case study in the potential pitfalls of a defined contribution plan. Since the mid-1960s, Nebraska provided all state and county workers with a defined contribution pension plan. After a large-scale study of the state pension system, the Nebraska state government changed its view of the DC plan. During the period from 1983 to 1999, state and county workers averaged 6 percent return on the investments in their individual accounts, whereas the state's traditional defined benefit plans achieved an average return of 11 percent (Sostek, 2004). Even an extensive educational program did little to improve participants' financial literacy. Due to the study, state lawmakers passed a bill in April 2002 abandoning the defined contribution plan and establishing a cash balance plan. Effective January 1, 2003, new employees had to participate in the newly created cash balance plan. As always is the case in the public sector, current employees were given the option to stay in the DC plan or convert to the CB plan. The basic features of the Nebraska CB plan are (Nebraska Public Employees' Retirement System, 2007):

1. The member of the plan contributes 4.8 percent of compensation each payroll. The state matches the member's contributions at the rate of 156 percent. There is a three-year vesting requirement for employer contributions.

2. The interest credit rate is defined as the greater of 5 percent, or the applicable federal midterm rate plus 1.5 percent. The federal midterm rate is based on the average market yield on outstanding marketable obligations of the United States with maturities of at least three years, but no more than nine years. For example, if the federal midterm rate is 4 percent, then the member will receive a 5.5 percent return on his account for that period. The interest credit rate is to be determined each calendar quarter (January, April, July, and October) based on the federal midterm rate that is published by the Internal Revenue Service as of the first day of that quarter. The rate is to be compounded annually.
3. All member and employer contributions are held in a trust fund, which is invested by professional fund managers under the direction of the Nebraska Investment Council. Trust fund dollars cannot be used for any purpose other than providing retirement benefits to members or covering plan expenses. When investment performance results are above the interest credit rate, the Public Employees Retirement Board (PERB) has the option to issue a dividend to active plan members, or the Nebraska Legislature may pass legislation to improve plan benefits for active members.
4. The normal form of payment under the cash balance benefit is a single life annuity with five-year certain, payable monthly. Members will have the option to convert their member cash balance account to a monthly annuity with built-in cost-of-living adjustments of 2.5 percent annually. This monthly benefit and all other options allowed under the plan will be of actuarial equivalence to the accumulated balance of the individual account. In addition to this basic annuity form, the Nebraska CB plan also offers other traditional annuities, such as joint annuity, period certain annuity, and life annuity with period certain.

The cash balance credit rate for 2003 through 2007 was 5.04, 5.19, 5.45, 6.27, and 6.12 percent, respectively, and due to the strong stock market performance during this period, the plan also distributed a dividend credit of 3.1, 2.8, 13.5, and 2.73 percent for years 2003 through 2006 (Nebraska Public Employees' Retirement System). To further reduce the funding risk to the retirement system, the system's board established a policy that sets a cap on the dividend. Effective 2007, all dividends granted, plus the annual interest credit, cannot exceed 8 percent unless a majority of the board agrees. By setting a minimum and maximum annual investment return on the members' notional account, the Nebraska's cash balance plan ensures a more stable and predictable long-term asset growth rate for its members.

7.4.6 California Cash Balance Plan

Another relatively small cash balance plan at the state level is for part-time employees of California's public schools (California State Teachers' Retirement System). The basic features of the California CB plan are as follows:

1. Each employer contributes a minimum of 4 percent of employee salary and the sum of the employee and employer contributions has to equal or exceed 8 percent of employee salary. A participant has an immediate vested right to employer contribution.
2. The guaranteed interest rate, determined annually by the Teachers' Retirement Board, is based on the average of 30-year U.S. Treasury notes for the 12 months immediately preceding the plan year.
3. A gain and loss reserve is established within the plan. The purpose of this reserve is to accumulate investment earnings above the guaranteed interest rate so that during the years when the actual return is below the guaranteed rate, funds can be withdrawn from the reserve to fully credit employee accounts.
4. After the end of the plan year, when the total investment earnings for the immediately preceding plan year are known, the board may also elect to give an additional earnings credit to employee accounts.
5. When the participant retires, he can claim his normal retirement benefit in two ways. He can take a lump-sum benefit equal to the balance of credits in his account. All of the lump-sum payment may also be eligible to roll over into an individual retirement account (IRA), defined contribution plan, or other eligible retirement plan that accepts such a rollover. The retiree can also opt to take one of the five annuity options: (1) a single life annuity with a cash refund feature, (2) a single life annuity without a cash refund feature, (3) a 100 percent joint and survivor annuity, (4) a 50 percent joint and survivor annuity, or (5) a period certain annuity.

7.5 Defined Contribution Plan Design Issues

When a government employer decides to switch from a defined benefit plan to a defined contribution plan, then several major DC plan design issues need to be considered carefully before implementation:

1. To whom will the new DC plan apply? As discussed many times in this book, pension rights are typically protected by state constitutions, statutes, and court decisions. Therefore, legally, a new DC plan can be made mandatory only for future employees and optional for current employees. If the majority of current employees opt not to participate in the new DC plan, then any major savings in pension cost, if this is the purpose of switching, will not be realized for several decades. If the pension plan has a significant unfunded liability, a switch to a DC plan will have minimal effect on its current liability. One of the most difficult decisions is to estimate the percentage of current employees who will switch to the DC plan. A wild card risk in such estimation is the financial market itself. A rising or a falling stock market can have a significant impact on the employees' decision to switch. This, in turn, will have an impact on the long-term cost of this switch to the government employer.
2. Even for new employees, a decision needs to be made whether the participation in the defined contribution plan should be mandatory or optional. If it is optional, an estimate needs to be made as to the percentage of them joining the DC plan and the impact on the long-term pension cost.
3. For current employees who opt to join the defined contribution plan, a decision needs to be made as to whether the accrued benefits of these employees should remain in the defined benefit plan or be transferred to the DC plan. If the accrued benefits are allowed to be transferred to the DC plan, then the next step is to decide what method should be used in valuing the accrued benefits.
4. The level of contribution from both the employee and employer is also a critical element of the decision process. Inevitably, the employer contribution rate in a DC plan will be compared to that in the current defined benefit plan.
5. The defined contribution plan sponsor also needs to decide whether there should be any vesting requirement, and, if so, how long the vesting period should be. While there are DC plans that have no vesting requirement, the more common practice is to require a vesting period of three to five years.
6. One of the more difficult design issues is the treatment of other benefits, primarily disability benefit. There are two design issues. For new employees who just joined the defined contribution plan, it needs to be decided whether they will be offered any disability benefit. For current employees who switch to a defined contribution plan, the question is what to do with the disability benefit accrued under the defined benefit plan. If employees in a defined contribution plan are not offered any disability benefit, then they may have to purchase disability insurance from a commercial insurance company. Florida offers an example of how to provide disability benefits to those in the defined contribution plan (Florida Retirement System, 2005). The Division of Retirement administers the disability program for the defined contribution plan. If a defined contribution plan member wants to receive disability benefit, he must apply for the benefit with the division. The eligibility requirements for disability benefits are the same as those for the members of Florida's defined

benefit plan. Service in the defined contribution plan will count toward the creditable service necessary for the DC member to vest for disability benefit. However, in order to receive the disability benefit, the DC plan member must transfer all the funds accumulated in his DC plan account to the Florida Retirement System.

7. Another critical defined contribution design issue is the investment provision. As long-term investment performance poses a key risk to DC plan members, plan sponsors will need to determine permissible investment options that will be allowed in the plan.

Of all the major defined contribution plan design issues facing the plan sponsor, the issue that will have the most impact on the long-term financial security of the majority of the people in a DC plan is the investment provision. What kind of investment options the plan sponsor allows the participant to choose from can play a critical role in determining the long-term investment performance. The perspective on investment options in DC plans has undergone major changes over time. Since DC plans allow the members to have total discretion over their investment, logically they should be offered as many investment options as possible. However, offering many investment options can be confusing and bewildering to many plan participants, especially if they are not well educated about investment. Hence, many options make it more likely that some of the participants will choose investment options that may not be suitable for their investment goals. When it comes to long-term investment, more options do not necessarily mean better performance for individual retirement accounts. For most people, limiting the investment offering to just a few simple and well-diversified funds will make it much easier for them to understand and result in a lesser chance of making grave long-term investment mistakes. When such thinking is pushed to the limit, then only one option needs to be offered to the plan member and all discretion over investment is taken out of the equation. It is such thinking that has led to the creation of lifecycle funds. The idea is to base the asset allocation of investment portfolio on the final retirement date. When a person is young, he has a longer investment horizon and, thus, can afford to be more aggressive (i.e., taking more investment risk) in his investment strategy. When he is getting closer to retirement, the investment strategy becomes increasingly more conservative by decreasing the allocation to risky assets (such as equity) and increasing the allocation to safe assets (such as fixed-income securities.) All of these shifts in asset allocation prior to retirement are done automatically by the managers of lifecycle funds with no member involvement. For example, if a member plans to retire in 2030, then on the first day he joins the plan, he can choose a lifecycle 2030 fund. His investment program will then be put on autopilot until he retires in 2030. The best advantage of a lifecycle fund is that for those with very little financial knowledge and for those who do not have the time to contemplate on all the investment options and strategies periodically, such an investment option will do all the analytical work for them and eliminate the chance of making grave mistakes that will severely damage long-term investment performance. Because of this advantage, lifecycle funds have become increasingly popular. In 1997, only $12 billion was invested in lifecycle funds; by 2006, $303 billion was invested in such funds (Investment Company Institute, 2006).

The federal Thrift Savings Plan (TSP) typifies the simple approach to long-term investment. The TSP is a defined contribution plan for civilians who are employed by the United States Government and members of the uniformed services. Congress established the TSP through the Federal Employees' Retirement System Act of 1986. In about 20 years, the TSP has become one of the largest defined contribution plans in the United States. By November 2005, there were almost 3.5 million participants and the total assets in TSP were valued at $171 billion (Thrift Savings Plan, 2006). One of the most unique characteristics of the TSP is that despite its enormous size, the investment options for the participants are both simple and limited. In the first year of the TSP's creation, there were only three funds available for investment: Government Securities Fund (G), Fixed Income Index Fund (F), and Common Stock Index Fund (C). The G Fund is invested in short-term U.S. Treasury securities specially issued to the TSP. The F Fund is to match the performance of the Lehman Brothers U.S. Aggregate Index, representing the broad U.S. bond market. The C Fund is to match the performance of the S&P 500 Index. These three very broad-based funds essentially capture the large-cap U.S. stock, bond, and cash-equivalent markets. It was not until 2001 when two more equity funds were added to the mix: Small Capitalization Stock Index Fund (S) and International Stock Index Fund (I). The S Fund is to match the performance of the Dow Jones Wilshire 4500 Completion (DJW 4500) Index, a broad market index made up of stocks of U.S. companies not included in the S&P 500 Index. The I Fund is to match the performance of the Morgan Stanley Capital International EAFE (Europe, Australia, and Far East) Index. By including S and I Funds, the TSP has captured the part of the equity market that was missing in the original options and, thus, provided further diversification. Table 7.2 shows the return on these five funds since inception and their benchmark performance (U.S. Government Thrift Savings Plan).

Table 7.2 TSP Funds Return (as of December 2005)

Fund	*G Fund*	*F Fund*	*C Fund*	*S Fund*	*I Fund*
Inception date	4/1/87	1/29/88	1/29/88	5/1/01	5/1/01
Return since inception[a]	6.59%	7.41%	11.64%	8.94	6.42
Benchmark return		7.78	11.96	9.04	6.71

[a] All returns are after expenses.

Source: TSP Fund information sheet.

The table shows that the long-term, after-expense returns of the funds closely track the benchmark returns. Even a portfolio consisting entirely of F Funds since its inception produced a respectable annual return by 2005. To make the investment decision even simpler and as the popularity of lifecycle funds has increased due to the advantage discussed earlier, the TSP added the Lifecycle Fund (L) to the mix in 2005. As of 2006, there were four L Funds, with target retirement years of 2010, 2020, 2030, and 2040. Each lifecycle fund consists of a different mix of the five funds already offered by the TSP. For example, the G Fund (the cash-equivalent fund) accounts for almost 50 percent of the Lifecycle 2010 Fund, whereas it is almost nonexistent for the Lifecycle 2040 Fund.

The case of the TSP demonstrates an elegantly simple and effective way to set up an investment program for a defined contribution plan. Lifecycle funds can become a core component of any defined contribution plan investment program that can also include a few broadly diversified funds. To make the investment decision even simpler for DC plan participants, it is also conceivable to make the lifecycle fund a default investment option for participants in a DC plan.

7.6 Conclusion

Since the stock market downturn in 2000, there has been a severe backlash against DB plans in the public sector due to the increasing cost and unfunded liability. There have been calls to shift to DC plans, a trend that has been witnessed in the private sector for quite some time. After many years of experience with DC plans, it has been found that DC plans also have their own risks. To address the funding risks of defined benefit plans to sponsors and those to participants in defined contribution plans, a third hybrid type of pension plan, the cash balance plan, has been introduced. When structured properly, CB plans can eliminate funding risks for both defined benefit and defined contribution plans as well as make pension benefits more portable compared to DB plans and pension assets less costly to manage compared to DC plans. While CB plans have been adopted by many employers in the corporate sector in the United States, its use in the public sector is very limited. Given the interest in DC plans in the public sector, CB retirement plans can also be considered as an alternative to traditional DB plans. In the absence of CB plans, more consideration should be given to designing an investment program that will help individual participants in defined contribution plans achieve an adequate long-term investment return.

References

Board of Trustees, Federal Old Age and Survivors Insurance and Federal Disability Insurance Trust Funds. 2004. *2004 Annual Report of the Board of Trustees of the Federal Old Age and Survivors Insurance and Disability Trust Funds* (Washington, DC: U.S. Government Printing Office).

California Public Employee Retirement System. 2006. *Comprehensive Annual Financial Report for the Fiscal Year Ended June 30, 2006.* https://www.calpers.ca.gov/mss-publication/pdf/x2N8fznMvFGq5_CAFR2006.pdf (Accessed April 10, 2007.)

California State Teachers' Retirement System. *Cash Balance Benefit Program Summary.* http://www.calstrs.com/Members/Cash%20Balance%20Benefit%20Program/summary.aspx (Accessed July 7, 2007.)

Florida Retirement System. 2005. *FRS Pension & Investment Plans Member Guide: Disability Benefits.* http://www.frs.state.fl.us/ (Access July 4, 2006.)

Gosselin, Peter G. 2005. States' Private Pensions Make a Weak Showing: the retirement accounts have had less appeal and spottier success than Bush plan's projections. *Los Angeles Times*, February 22, 2005, A1.

Indiana Public Employees' Retirement Fund. *Member Handbook.* http://www.in.gov/perf/pdf/public/member_handbook_printable.pdf (Accessed June 14, 2007.)

Investment Company Institute. 2006. *Research Fundamental: The U.S. Retirement Market, 2006.* Vol. 16, No. 3. http://www.ici.org (Accessed December 20, 2006.)

Munnell, Alicia H., Mauricio Soto, Jerilyn Libby, and John Princivalli. 2006. *Investment Returns: Defined Benefits vs. 401(k) Plans.* Working Paper. Center for Retirement Research, Boston College. http://www.bc.edu/centers/crr/issues/ib_52.pdf (Accessed December 19, 2006.)

Nebraska Public Employees' Retirement Systems. *State Employees Retirement System.* http://www.npers.ne.gov/public/howto/handbooks/handbookState.pdf (Accessed August 8, 2007.

Nebraska Public Employees' Retirement Systems. *Cash Balance Rates of Return and Dividends.* http://www.npers.ne.gov/public/aboutus/CashBalInfo.jsp (Accessed December 13, 2007.)

Ohio Public Employees Retirement System. *Plan Comparison Chart.* https://www.opers.org/members/plan-select/tools/chart.shtml (Accessed September 18, 2007.)

Ohio Public Employees Retirement System. 2007. *The Benefits of Membership: a Comprehensive Overview.* https://www.opers.org/pubs-archive/members/member-handbook.pdf#zoom=100 (Accessed April 10, 2008.)

Oregon Public Employees Retirement System. *What is OPSRP?* http://www.oregon.gov/PERS/MEM/section/general_information/what_is_opsrp.shtml (Accessed September 18, 2007.)

Sostek, Anya. 2004. Pension pendulum. *Governing Magazine,* March, 2004.

State of Michigan, Department of Management and Budget, Office of Retirement Services. *401(k) Retirement Plan Overview.* http://www.michigan.gov/orsstatedc/0,1607,7-209-34551---,00.html (Accessed August 7, 2007.)

State of Michigan, Office of Retirement Services. 2004. The 401(k) Retirement Plan for State of Michigan Employees. http://www.michigan.gov/documents/DC1_88666_7.pdf (Accessed April 10, 2008.)

State of Washington, Department of Retirement Systems. *PERS Member Handbook.* http://www.drs.wa.gov/member/Publications/PERS/handbooks/Pers3/pers3hbk.pdf (Accessed May 15, 2007.)

Thrift Savings Plan. *TSP Fund Information Sheet.* http://tsp.gov/rates/fundsheets.html (Accessed December 11, 2006.)

U.S. Governmental Accountability Office. 2000. *Cash Balance Plans: Implications for Retirement Income* (HEHS-00-207), Washington, D.C.

U.S. Government Thrift Savings Plan. TSP Fund Information Sheet. http://tsp.gov/rates/fundsheets.html (Accessed December 11, 2006.)

U.S. Government Thrift Savings Plan. 2006. *Thrift Savings Plan: Highlight.* http://tsp.gov/forms/high06a.pdf (Accessed December 21, 2006.)

Chapter 8

Other Postemployment Benefits

Besides traditional pension benefits, retirees in the public sector also enjoy many other postemployment benefits (OPEB). While healthcare subsidy remains the largest part of OPEB by far, OPEB can also include dental and vision care as well as life, disability, and long-term care insurance. Since OPEB is also tied to employment, it shares one important characteristic with pension benefits, namely the benefit is earned during one's working years, but will not be paid until after retirement. OPEB, therefore, should be funded and reported in a way that is similar to pension benefit, meaning OPEB costs should be recognized and funded in the period when it is earned by the employee rather than when it is paid to her during retirement. While the Government Accounting Standards Board (GASB) issued Statements 25 and 27 in 1994 to set the standards for financial reporting on pension benefits that require actuarially determined information, no such information was required of financial reporting on OPEB, which was administered on a pay-as-you-go basis in most states. In 2004, GASB issued two statements: Statement 43, *Financial Reporting for Postemployment Benefit Plans Other Than Pension Plans,* and Statement 45, *Accounting and Financial Reporting by Employers for Postemployment Benefits Other Than Pensions,* to set standards for financial reporting on OPEB. These two statements are substantially similar to Statements 25 and 27 in terms of reporting standards.

The issuance of GASB Statements 43 and 45 serves at least three purposes. First, with the recognition of the cost of OPEB in the period it is earned, the true cost of personnel and, thus, government services becomes more transparent. Second, with information on the unfunded actuarial accrued liabilities (AALs) for promised OPEB associated with past service, there is also a better recognition of the financial impact of government commitments made in the past and the full extent of government financial liability. Third, if such recognition leads to prefunding of OPEB costs on the part of the government employer or reduction in OPEB to retirees, the principle of intergenerational equity will be met in the financing of public services and the financial burden on future generations of past commitments will be lessened.

In the first section of this chapter, we look at what healthcare for retirees involves. In the second section, we examine the financial reporting standards in GASB 43 and 45 as well as the financial implication of these two statements. In the final section, we analyze how state and local governments deal with the OPEB liability.

8.1 Healthcare for Retirees

As in the case of pension benefits, there are also two different types of healthcare benefit plans for retirees: defined benefit and defined contribution. A defined benefit OPEB plan specifies the amount of benefits to be provided, either in dollars or at a certain level of coverage (such as paying for a certain percentage of healthcare insurance premiums.) In a defined contribution plan, the plan provides an individual account for each member and the employer makes periodic contributions to the individual account, with the contribution being a set dollar amount or a percentage of payroll. After the member retires, the plan sponsor is no longer responsible for her health benefit. The benefit depends on the balance of the individual account, which is a combination of contributions and investment earnings. Just as in the case of pension benefits, the defined benefit OPEB plan is the predominant plan type in the public sector.

Since people over 65 years old are covered by Medicare and people in the public sector retire at an age earlier than that on average, heath benefits for public retirees can be divided into two categories, pre-Medicare health benefits and Medicare-eligible health benefits. According to a survey by the American Association of Retired Persons (AARP), in fiscal year 2003, all 50 state government employers surveyed offered healthcare benefits for retirees under the age of 65, and all but Indiana and Nebraska offered healthcare benefits to retirees age 65 and older (Wisniewski and Wisniewski, 2004).* The level of health benefits offered to public sector retirees varies significantly among state and local jurisdictions. There are several major aspects of healthcare benefit design that determine the level of such benefits to retirees. First, there is an eligibility requirement for health benefits, the same as that for pension benefits in terms of years of service and age. Some states make it more restrictive for retirees to receive health benefits by imposing a longer service requirement.

* The author would like to thank the Public Policy Institute of AARP for giving permission to use the data from this research study sponsored by the Public Policy Institute.

According to the same AARP survey, 9 states require retirees to have 10 or more years of service to become eligible for health benefit.

Second, and more importantly, the amount retirees have to pay for the health benefit, in terms of healthcare premium and out-of-pocket payment, varies significantly from one state to the other. Health benefit is primarily in the form of health insurance premium subsidy from the employer. According to the AARP survey, 16 states paid the full amount of the premium for at least one of the health plans offered to the retirees; in 12 states, the retiree paid 100 percent of the premium; in the rest of the states, retirees paid a portion of the premium. Some states pay the premium on a sliding scale based on years of service, with a lower subsidy rate for retirees with fewer years of service. In addition to the variation in premium subsidy rate, there is also substantial variation in the premium itself, ranging from a low of $159.92 in South Dakota to a high of $925.42 for a PPO (preferred provider organization) plan in Arizona.

States also vary in the amount of out-of-pocket payment on the part of retirees. Such payment includes copayment for doctor visits, deductible, and prescription drug cost. Most states also set a maximum out-of-pocket expense to protect the retirees. Deductible varies from nothing to over $1000, and maximum out-of-pocket can also vary from a few hundred dollars in some states to $10,000 in Wyoming.

Due to the differences in the size of healthcare premiums, premium subsidy rate, and out-of-pocket expenses, there will be a substantial difference in the magnitude of OPEB liability among states on a per-employee basis.

8.2 GASB Statements 43 and 45

GASB Statements 43 and 45 are the culmination of a long-running effort by GASB to provide standards on financial reporting for OPEB. Before reviewing the reporting standards contained in these two statements, it is worthwhile to review previous GASB efforts leading up to them.

8.2.1 Brief History of GASB Statements on OPEB Reporting

GASB efforts on OPEB financial reporting date back to 1990, when it issued Statement 12, *Disclosure of Information on Postemployment Benefits Other Than Pension Benefits by State and Local Governmental Employers*. The issuance of GASB 12 was in conjunction with the Financial Accounting Standard Board (FASB) issuing Statement No. 106, *Employers' Accounting for Postretirement Benefits Other Than Pensions*, which provided standards for OPEB financial reporting in the private sector. GASB 12 requires that all governmental employers who finance all or some portion of their retiree health benefit costs should provide, among other things, a description of the benefits provided and employee groups covered, and the expenditures/expenses for those benefits recognized for the period. Statement No. 12 did not require the calculation of an AAL or the recognition of current employees' accrued benefits.

In 1996, GASB issued Statement 26, *Financial Reporting for Postemployment Healthcare Plans Administered by Defined Benefit Pension Plans.* Under GASB 26, retiree healthcare benefit plans are required to present a statement of plan net assets, a statement of changes in net assets, and note disclosures similar to those required of pension plans. However, GASB 26 does not require that these retiree health plans provide the "required supplementary information" applicable to pension plans, i.e., a schedule of funding progress, a schedule of employer contributions, and related notes, such as valuation methodology and key assumptions employed in the valuation.

8.2.2 GASB Statement 43

GASB 43 establishes standards for financial reporting of OPEB plans, which are trust or other funds that hold assets to fund OPEB. The financial reporting standards are substantially similar to those for pension plans set in GASB 25, which were discussed in Chapter 3, with modifications to reflect some differences between pension plans and OPEB plans.

GASB 43 makes an important distinction between those OPEB plans set up as trust funds and those set up as nontrust funds. For a plan to be qualified as an irrevocable trust, three standards have to be met: (1) employer contributions to the plan are irrevocable, (2) plan assets are dedicated to providing benefits to their retirees and beneficiaries, and (3) plan assets are legally protected from creditors of the employer(s) or the plan administrator.* Any other arrangement, such as a separate governmental fund in which assets are accumulated for OPEB purpose, is not considered a trust. While the standards in GASB 43 apply to the financial reporting for OPEB plans held in trust, OPEB plans not administered as a trust also have to disclose certain information required of OPEB trust funds. The major implication of the distinction between a trust and nontrust fund, however, is in the way that OPEB liability and annual required contribution are determined, as will be discussed later in this section.

The standards for financial reporting fall into two main categories: framework for reporting and parameters for actuarially determined information.

* GASB 43, Paragraph 4.

8.2.2.1 Financial Reporting Framework

The financial reporting framework for defined benefit OPEB plans that are administered as trusts are very similar to those for defined benefit pension plans: two financial statements, notes to the financial statements, and two schedules presented as required supplemental information (RSI).*

The two financial statements are a statement of net assets and a statement of changes in net assets. They focus on the current financial situation of the plan. The statement of net assets provides information on the fair market value of plan assets and plan liabilities. The statement of changes in plan net assets provides information on financial activities that cause the net assets to change from year to year. The main activities to increase net assets are contribution and investment income, and the main activities that decrease net assets are benefit payment, contribution refunding, and administrative expenses. Notes to the financial statements include a brief plan description (such as benefits offered and the eligibility for benefits), a summary of significant accounting policies, and funding policy.

The two required schedules are schedule of funding progress and schedule of employer contributions. The schedule of funding progress provides information on the actuarial value of assets, the actuarial accrued liability, and the funding ratio of the plan. The schedule of employer contributions reports the annual required contributions (ARC) of the employer and the percentage of ARC being made by the employer. As in the case of pension plans, these two schedules, with their historical trend information, provide a long-term perspective on the financial situation of the plan. The required schedules are also followed by notes, disclosing information on the actuarial cost methods and assumptions used in arriving at the information contained in the schedules.

8.2.2.2 Parameters

Plans are required to measure all actuarially determined information included in their financial reports in accordance with certain parameters. The key parameters include actuarial methods, amortization, and assumptions. If the methods and assumptions used in determining a plan's funding requirements meet the parameters, the same methods and assumptions are required for financial reporting by both a plan and its participating employer(s). However, if a plan's funding policy does not meet the parameters, such as a pay-as-you-go funding policy, the parameters still apply for financial reporting purposes. Following is a discussion of some of the main parameters.

Actuarial cost method — The six actuarial cost methods permissible under GASB 43 are the same as those used to determine pension liability under GASB 25: entry age normal (EAN), frozen entry age, attained age, frozen attained age,

* GASB 43, Paragraph 17.

projected unit credit, and aggregate cost.* GASB, however, also singles out the aggregate cost method for further clarification. Because this cost method does not separately calculate unfunded liability, it does not yield a funded ratio. Since most of them are not actuarially funded and have little or no assets, an aggregate cost method will provide little information about the financial condition of the OPEB plan. Therefore, GASB 43 requires plans that select an aggregate cost method for funding purpose also must prepare a schedule of funding progress for financial reporting purpose by using the EAN cost method.†

Amortization — GASB 43 sets the maximum amortization period to be thirty years and, if different parts of unfunded liability are amortized separately, then the single equivalent amortization period can not exceed thirty years.‡ Like amortization for unfunded pension liability, the amortization period can be both closed and open, and the amortization amount can be either level dollar or level percentage of payroll. Any actuarial gains as a result of switching to a new actuarial cost method, however, have to be amortized over a minimum of eleven years.

Assumptions — While many economic and demographic assumptions used in determining OPEB liabilities are similar to those for pension liability, such as investment return, inflation rate, mortality rate, and turnover rate, there is a major difference in the determination of the assumed rate of return between the OPEB plan and the pension plan. While all pension plans use the long-term expected rate of return as the assumed rate, the situation is more complicated for OPEB plans due to the fact that very few governments have accumulated any assets for OPEB plans. Depending on the funding situation, one of three rates will be used as a discount rate. First, if an OPEB plan is set up as an irrevocable trust and the employer fully contributes the ARC, the assets are considered plan assets and a long-term rate of return can be used as the discount rate,§ similar to that used by a pension plan.¶ Second, if an OPEB plan is not set up as an irrevocable trust, then assets in the plan are considered employer assets. Since government employer assets are usually limited to short-term investment instruments, the discount rate in this arrangement is also limited to short-term investment return, which is much lower than long-term investment return.** Third, for an irrevocable trust to which the employer contributes less than the required amount, then a blended rate is used as the discount rate. The blend rate, a mix of long-term and short-term rate of returns, will be lower than the long-term rate, but higher than the short-term rate. While GASB 43 does not specify a method to determine a blended rate, it suggests several approaches, such as funded ratio approach (based on the extent to which a

---

* GASB 43, Paragraph 34.d.

† GASB 43, Paragraph 111.

‡ GASB 43, Paragraph 34.f.

§ GASB 43, Paragraph 138.

¶ GASB 43, Paragraph 34.c.

** GASB 43, Paragraph 136.

plan is funded) and the ARC approach (based on the percentage of the ARC actually being contributed).*

Even though the purpose of using three different kinds of investment return assumptions is to better reflect what type of investment return can be reasonably expected in the long run, the choice of discount rate also has major implications for calculating OPEB liability and the annual required amount. Since all future OPEB costs are discounted to determine the present value of OPEB liability, everything being equal, a higher discount rate results in a smaller present value of liability and, thus, higher funded ratio. The ARC, consisting of normal cost and amortization cost, will also be smaller with a higher discount rate, everything else being equal. This means that OPEB plans set up as a trust will have smaller liability and smaller annual cost compared to those plans not set up as a trust; and employers who contribute the required amount regularly to a trust fund will also face lower costs than those who do not fully pay the required trust fund amount. Therefore, the existence of three discount rates provides an incentive for employers to prefund OPEB costs and meet the financial obligation every year.

8.2.2.3 Implicit Subsidy Rate

GASB 43 requires that when an employer provides benefits to both active employees and retirees through the same plan, the retiree benefits should be segregated and measured independently. Further, future retiree benefits should be projected based on claims costs or on age-adjusted premiums approximating claims costs, rather than on unadjusted premiums.† This deals with the issue of implicit rate subsidy. One reason postemployment healthcare is so expensive is that healthcare costs tend to increase significantly with age. Premiums for retirees normally should be significantly higher than premiums for active employees. It is a common practice for employers to permit retired employees to continue in the employer's group health insurance plan, which also covers active employees, by paying the same premium charged to active employees. Thus, active employees are overcharged to subsidize retirees' healthcare costs. This creates an implicit rate subsidy, which is considered an OPEB benefit that should be included in OPEB valuations. The consequence of recognizing this implicit subsidy rate is that even if retirees pay 100 percent of the premium for their healthcare insurance, this implicit rate subsidy will force an employer that does not pay for retirees' premiums to recognize OPEB liability.

---

* GASB 43, Paragraph 138.

† GASB 43, Paragraph 34.a (2).

8.2.3 GASB Statement 45

GASB 45 establishes the standards for financial reporting of OPEB for state and local government employers. It includes most of the standards and parameters already addressed in GASB 43. The most important part of GASB 45 centers on the discussion of government employer's annual OPEB cost and its net OPEB obligation. Employers are required to measure and disclose annual OPEB cost on the accrual basis of accounting. Annual OPEB cost is equal to the employer's annual required contribution to the plan, with certain adjustments if the employer has a net OPEB obligation. An employer's net OPEB obligation is defined as the cumulative difference between annual OPEB cost and the employer's contributions to a plan.* If an employer pays the full amount of the ARC every year, the employer will not have an OPEB obligation and the annual OPEB cost is equal to ARC. However, when an employer contributes less than the ARC, a net OPEB obligation is created. When this occurs, the calculation of ARCs in subsequent years must include an amount for amortization of past contribution deficiency, and the annual OPEB cost will no longer be equal to ARC.

The calculation of annual OPEB cost involves two adjustments to the ARC: (1) the ARC is increased by the lost interest earnings on amounts not contributed and (2) decreased by the amount included in any subsequent calculation of the ARC to recapture, or amortize, the prior contribution deficiency.† These adjustments prevent OPEB costs that were recognized previously from being recognized again. For the same logic, when there is a negative OPEB obligation, meaning an employer contributes more than the ARC, the annual OPEB cost in subsequent years will be decreased by the interest earnings on amounts contributed in excess of ARC and increased by the amount included in any subsequent calculation of the ARC to recapture, or amortize, the excess contribution.‡

There is one major consequence of having a net OPEB obligation. GASB 45 requires that positive (or negative) net OPEB obligations be displayed as liabilities (or assets) in government-wide financial statements.§ Thus, a positive net OPEB obligation will have a very visible negative impact on the government employer's overall financial situation.

8.2.4 Financial Implication of GASB Statements 43 and 45

Due to the fact that most state and local governments have not set aside any assets for future retiree health costs, the unfunded OPEB liability is substantial. Following are some states' OPEB liabilities:

---

* GASB 45, Paragraph 14.

† GASB 45, Paragraph 15.

‡ Ibid.

§ GASB 45, Paragraph 21.

- California: $31 to $48 billion, depending on future funding method (Gabriel, Roeder Smith, & Company, 2007)
- Connecticut: $21 billion (State of Connecticut, 2006)
- New Jersey: $31 to $58 billion, depending on funding method (Aon Consulting, 2007)
- New York: $47 billion (Office of the New York State Comptroller, 2007)

Using available data on OPEB liability and extrapolation, two separate studies in 2006 and 2007 found that the public sector's total OPEB liability was between $1.4 and $1.5 trillion (Edwards and Gokhale, 2006; Zion and Varshney, 2007). At this level, the public sector's total unfunded OPEB liability dwarfs its unfunded pension liability.

The magnitude of unfunded OPEB liability clearly indicates the fiscal challenge facing state and local governments and gives rise to two questions: (1) how does this unfunded liability affect government credit rating and (2) how do state and local governments deal with this liability? In terms of the first question, all three major rating agencies, Moody's, Standard and Poor's, and Fitch, share fairly similar views on the credit impact of OPEB liability: While the absolute level of OPEB liability does not have an immediate impact on credit rating, the government's actions, such as designing a plan to reduce the unfunded liability and meeting its future OPEB funding obligation, are important factors in the assessment of a government's credit quality (Moody's Investors Service, 2005; Standard and Poor's, 2004; Fitch Ratings, 2005).

This view of rating agencies also dovetails with the reporting requirement of GASB 45. GASB 45 does not require a state or local government to fund the OPEB benefits on an actuarial basis, but the cumulative difference between the pay-as-you-go costs and the annual required contributions must be reported and disclosed on the government-wide financial statements as a liability. If the government employer does not have a plan to reduce this liability, there will be a growing net OPEB obligation reported in the financial statement. This will weaken the government's overall financial condition and will have a negative impact on the government's credit rating.

The answer to the second question is addressed in the next section.

8.3 Dealing with OPEB Liability

Given the magnitude of the OPEB liability and its implication on credit ratings, the challenge facing state and local governments is to devise a plan to gradually reduce the liability. There are three basic strategies in dealing with the escalation in OPEB costs: prefunding OPEB costs, reducing health benefits for future retirees, and switching from a defined benefit health plan to a defined contribution health plan. Even though these three methods are discussed separately, they are not mutually exclusive. For example, prefunding and reduction in future health benefits can be part of an overall strategy in reducing future OPEB liability.

8.3.1 Prefunding OPEB Liability

Prefunding means OPEB liability will be funded on an actuarial basis, and the annual OPEB cost includes both normal cost and another amount to amortize the unfunded liability. This requires the establishment of an OPEB trust fund, into which an employer and employees make periodic payments to fund future OPEB obligations.

Prefunding OPEB liability is by no means a new concept in the public sector. Some governments have been prefunding OPEB for some time. In the survey conducted by AARP, eleven states were found to be prefunding OPEB in 2001 (Wisniewski and Wisniewski, 2004). Some local governments have also started prefunding OPEB since the 1980s. Ohio's Postemployment Healthcare Plan (PHCP) is the largest OPEB trust fund in the public sector, with $11.1 billion in assets at the end of 2005 (State of Ohio, 2006). Most of the healthcare trust funds are administered within the public employee retirement systems. For example, Ohio's PHCP is a separate trust fund within the Ohio Public Employees Retirement System (OPERS). The PHCP assets are separate from the system's pension assets, although they can be pooled together for investment purposes.

8.3.1.1 OPEB Trust Fund

While prefunding will certainly entail higher OPEB costs initially when compared to the pay-as-you-go funding method, it generates many benefits in the long run for both employers and employees. Even though GASB 45 does not require employers to prefund OPEB liabilities, it provides strong incentives for government employers to set up an irrevocable trust fund to do so. This incentive, as discussed earlier, comes in the form of a higher discount rate used to calculate OPEB liability and annual OPEB contribution, resulting in smaller OPEB liability and ARC. In addition, prefunding has other benefits:

- Assets accumulated in the trust fund are invested and investment earnings will reduce employer OPEB costs in the long run.
- Prefunding leads to more stable OPEB payments over time rather than an escalation of payments under the pay-as-you-go funding situation, resulting in more stability for the operating budget in the long run.
- It will gradually eliminate the unfunded OPEB liability and, thus, reduce the expenses related to the unfunded liability and a gradually decreasing unfunded liability also improves the government's balance sheets and credit rating.

- Prefunding also improves the security of OPEB for retirees.

Given these benefits and the incentive in establishing an irrevocable trust fund to prefund OPEB liability, many state and local governments have already established or considered establishing such trust funds in the wake of GASB 43 and 45. Following are some examples of OPEB trust funds established after the publication of these two statements:

- **California:** In February 2007, the California Public Employees' Retirement System (CalPERS) announced the creation of a new trust fund, called the California Employers' Retiree Benefit Trust Fund that allows public employers to prefund the future OPEB liabilities. It allows participating employers to make regular periodic contributions into the trust fund. The contributions are invested by CalPERS to take advantage of its size and investment team.
- **Georgia**: The State of Georgia enacted a law in 2005 creating the Georgia Retiree Health Benefit Fund to receive annual contributions based on the state's actuarial required contribution, and assets in the fund can only be used to pay for fund obligations. In his budget proposal for fiscal year 2008, the governor proposed setting aside $100 million in the fund to pay down the OPEB liability of about $15 billion (McCaffrey, 2007).
- **New York City:** New York City's OPEB liability was over $53 billion. To pay down this debt, the city established the Retiree Health Insurance Benefit Trust. Due to the strong economy in New York City and the budget surplus, the city used part of the budget surplus to deposit $1 billion into the trust fund in each fiscal year in 2006 and 2007.

8.3.1.2 OPEB Bond

Establishing an irrevocable trust fund has another potential advantage over the pay-as-you-go funding method. While it is common to pay off the unfunded OPEB liability through amortization over a 30-year period, another method available to employers with an OPEB trust fund is to pay off part or all of the unfunded liability with an OPEB bond. An OPEB bond works in the same way as a pension obligation bond, which was discussed in Chapter 6. The proceeds of an OPEB bond are deposited into the trust fund to pay off the unfunded OPEB liability and the plan sponsor will then make debt service payments to investors in the OPEB bond. As an OPEB bond and a pension obligation bond work in essentially the same way, they also share the same benefits and risks.

On the benefit side, as the government employer also has to pay interest on the unfunded OPEB liability, if the OPEB bond can be issued at an interest rate below the discount rate for the OPEB liability, the government issuer can realize interest cost savings. Even if the return on the investment of OPEB bond proceeds fails to exceed the discount rate, the government will benefit from an OPEB bond as long as the investment return exceeds the interest rate on the bond. An additional benefit of an OPEB bond is that because the OPEB liability is much greater than unfunded pension liability and there will be few assets in a trust fund to begin with, an infusion of OPEB bond proceeds will significantly boost the OPEB plan funding ratio, allowing the plan sponsor to use a higher discount rate in a situation when a blended rate has to be used. This will lower the government employer's long-term OPEB costs. While debt service on OPEB bonds will generally be higher than pay-as-you-go financing costs for the first few years, pay-as-you-go costs are likely to increase sharply and exceed the cost of debt service after a few years, and the cost difference will continue to grow thereafter.

The main risks of an OPEB bond are investment risk and timing risk, similar to those for a pension obligation bond. If the actual investment return fails to exceed the interest rate on the OPEB bond, the total cost will be higher to the issuer than if the bond were not issued. The timing risk refers to the possibility of a substantial decline in the financial market soon after the proceeds of the bond are invested. This is more of a risk for OPEB bond than for pension obligation bonds (POBs), as OPEB bond proceeds account for a much greater percentage of the total assets in the trust fund. However, as in the case of POBs, the true benefit and cost of an OPEB bond can be known only over the entire life of the bond. Another disadvantage is the reduced financial flexibility. Since unfunded OPEB liability is fairly large, the debt service on an OPEB bond will also be fairly large relative to the government's operating budget, and, therefore, can constrain its financial flexibility during times of fiscal stress.

Some state and local government employers have already issued OPEB bonds. In 2003, the State of Wisconsin sold $1.8 billion in taxable pension bonds to pay off unfunded pension liability and unfunded OPEB liabilities (State of Wisconsin, 2004). The City of Gainesville, Florida, began funding its OPEB liability on an actuarial basis in 1994 through its Retiree Health Plan. By the end of 2004, the liability was approximately 23 percent funded (City of Gainesville). In 2005, the city issued a ten-year OPEB bond totaling $35 million to fully fund its retiree health plan.

8.3.2 Reducing Health Benefits

Reducing health benefits is another response for many governments to contain OPEB costs. Before benefit reduction can be implemented, a legal issue has to be resolved: whether health benefits are protected by law and cannot be reduced. While pension benefits are fully protected, as discussed in Chapter 4, the legal protection of health benefits is more ambiguous.

8.3.2.1 Legal Perspective

The protection of pension benefits comes from state constitutions, statutes, and court rulings. The court typically considers an employee earns the contractual right to pension benefits at the time of employment and, thus, pension benefits cannot be impaired under this contractual agreement. This concept of contractual right that underpins court decisions with regard to pension benefits is not automatically extended to OPEB benefits. According to a study commissioned by the National Conference on Public Employee Retirement Systems, there are 19 states that have court decisions regarding retiree health benefits.* These 19 states can be divided into three groups.

In the first group, there are seven states in which the courts decided that health benefits are not protected by contractual rights and retirees do not have vested rights in health benefits. The seven states are: Colorado, Iowa, Massachusetts, Michigan, Rhode Island, Pennsylvania, and Tennessee. In another state, the Maryland State Attorney General expressed the opinion that the state can change the laws regarding such health benefits. Michigan's case typified the court decisions in this group. In 2000, the state of Michigan made changes to the retiree healthcare plans by increasing copayments and out-of-pocket maximums for prescription drugs and health insurance deductibles for retirees. Six retired members of the Michigan Public School Employees' Retirement System (MPSERS) sued the Michigan Public School Employees' Retirement Board, the Michigan Public School Employees' Retirement System, the Department of Management and Budget, and the Treasurer of Michigan, alleging that such changes amounted to a diminishment and impairment of their accrued financial benefits under the state constitution. In 2005, the Michigan Supreme Court reached a decision that favored the state. The court decision contained two parts. First, it held that healthcare benefits do not constitute "accrued financial benefits" subject to protection of impairment or diminishment under the state constitution. Second, the court also held that Michigan's statute that established healthcare benefits was not intended to create a contractual relationship with public school employees and, thus, payment of healthcare benefits by the MPSERS is not a contractual right subject to protection under the state constitution and can be changed by later legislation.†

The second group contains seven states in which the courts decided that retiree health benefits have vested rights and, therefore, are protected from impairment. These seven states are: Alaska, California, Kentucky, Minnesota, New Jersey, Oklahoma, and Wisconsin. Alaska typifies the court decision in this group. In 1999 and

---

* National Conference on Public Employee Retirement Systems. *State Cases Addressing Public Sector Health Benefits.* http://ncpers.org/Files/News/03152007HealthBenefitProtections.pdf (Accessed August 13, 2007.) The author would like to thank NCPERS for giving the author the permission to use the material in the survey.

† Michigan Supreme Court. 2005. *Alberta Studier, et al., v. Michigan Public School Employees' Retirement Board, et al.*, No. 125765 and 125766, Lansing Michigan, June 28, 2005: 2–6.

2000, the state made changes to its group health insurance plan for retired public employees. Some of the changes led to a reduction in benefits, such as increases in the deductible and maximum out-of-pocket payments. In 2000, three retiree organizations and individual members brought separate lawsuits against the state, alleging such changes violated Alaska State Constitution by diminishing accrued benefits. In 2003, Alaska's Supreme Court reached the decision in favor of the plaintiff in *Duncan v. Retired Public Employees of Alaska*. The court held that the "accrued benefits" of retirement systems, which cannot be diminished and impaired under the state constitution, also include health insurance benefits, and, thus, such benefits are protected by the state constitution. The court also held that "the natural and ordinary meaning of 'benefits' in a health insurance context refers to the coverage provided rather than the cost of the insurance." At the same time, the court also held that "the prohibition on diminishment or impairment of retirement benefits does not mean that retirement benefits are unchangeable. Instead, benefits can be modified so long as the modifications are reasonable, and one condition of reasonableness is that disadvantageous changes must be offset by comparable new beneficial changes."*

The third group includes four states, Connecticut, Georgia, New York, and West Virginia, which protect retiree health benefits to some extent. For example, courts in Connecticut and Georgia held that while health benefits are vested and protected, government employers can change the form of health insurance, such as switching to managed care or health maintenance organizations (HMOs). New York state legislature enacted in 1994 "Retiree Healthcare Moratorium," which precludes any diminution of a retiree's health insurance benefits "unless a corresponding diminution of benefits" is applied to the corresponding group of active employees. A court in West Virginia reached a similar decision regarding health benefits.

This brief review of legal cases shows that retiree health benefits do not enjoy the universal protection given to pension benefits. It is up to individual states to decide, either statutorily or through court decision, to what extent such benefits are protected.

8.3.2.2 Reducing Health Benefits

Reduction in OPEB costs can be achieved through two means: either by having retirees paying a larger share of the healthcare costs or by limiting the eligibility to healthcare programs. Methods that shift a larger share of health costs to retirees include increases in retiree health premiums, higher copayments, higher deductibles, and higher caps on retirees' out-of-pocket expenses. For example, in 2006, Orange County, California, took several steps to reduce its health benefit costs. The

---

* *Duncan v. Retired Public Employees of Alaska, Inc.* 71 P.3d 882. Alaska, 2003. June 13, 2003.

most important step was to split retiree and active employee health plans and to require retirees to pay higher premiums (Orange County, California, 2006).

The method to limit eligibility to health benefit is to extend the vesting period for partial or full health benefit. Due to turnover rates, extending the vesting period essentially serves the purpose of eliminating health benefits for certain people and reducing benefits for others. Several state and local governments have used this method to contain OPEB costs. For example:

- In **North Carolina**, an actuarial study in 2005 showed that the state had an OPEB liability of about $24 billion at the end of 2005. The state paid 100 percent of health premiums for retirees who had at least five years of state service. To reduce future OPEB cost, the state changed the vesting requirement. The current legislation requires that employees hired after October 1, 2006, must work twenty years to receive noncontributory healthcare benefits in retirement. Employees with ten, but less than twenty, years of service, must pay 50 percent of the premium for retiree healthcare benefits. Employees with less than ten years of service will not receive retiree health benefits.
- In **Santa Clara County, California** (2006), the vesting requirement for employees hired before August 1996 is five years of service after attaining age 50; for those hired between August 1996 and June 2006, the vesting requirement is eight years after attaining age 50; and for those hired after June 2006, the requirement is ten years of service after attaining age 50.

Ohio state government has implemented a comprehensive reform of its healthcare program for retirees, including cost shifting, eligibility limitation, and giving retirees more health plan options. Even though Ohio's Postemployment Healthcare Plan is the largest healthcare trust fund in the public sector, its funding ratio was still relatively low at 35 percent at the end of 2005. To ensure the healthcare trust fund remains solvent for 15 to 25 years, the Ohio Public Employee Retirement System (OPERS) board in 2004 approved the Healthcare Preservation Plan (HCPP), which included the following main features:*

1. **Increase contribution**: Beginning in 2006, member contribution rates will increase by 0.5 percent each year for three years, increasing from 8.5 percent of payroll to 10 percent. Similarly, employer contribution rates will increase from 13.31 percent up to the statutory maximum of 14 percent over the three years beginning in 2006. Of this amount, the contribution rate to the healthcare trust fund would increase from 4 percent in 2004 to 5.5 percent in 2008.

* The information on HCPP is largely drawn from "Health Care Preservation Plan," by the Ohio Public Employees Retirement System, published in September 2004.

2. **Graded monthly allocation**: Retirees eligible for healthcare benefits will receive a graded monthly allocation (GMA) based on their years of service at retirement. A full allocation (100 percent) is equal to the amount that OPERS pays healthcare expenses for a retiree for one month. For employees hired before January 1, 2003, those with ten to fifteen years of service will receive 50 percent. This allocation will increase by 3.33 percent for every additional year of service until it reaches 100 percent with thirty years of service. For employees hired after January 1, 2003, those with ten to fifteen years of service will receive only 25 percent and each additional year afterward will increase the allocation by 5 percent, reaching 100 percent with thirty years of service.
3. **Cafeteria plan and retiree medical account (RMA)**: The HCPP features a cafeteria-style health plan. Retirees can choose from three different levels of coverage: Enhanced, Standard, and Basic. A lower level of coverage charges a smaller premium, but it also charges higher copayments, deductibles, and maximum out-of-pocket expense. A retiree can use his monthly allocation to purchase a health plan coverage option that best fits his own situation. If the monthly allocation exceeds the cost of the option the retiree selects, the balance will go into an RMA and can be used for future healthcare expenditures. Funds accumulated in an RMA can be rolled over from year to year. However, the balance of an RMA is nonrefundable and cannot be cashed out or borrowed against at any time.

8.3.3 Defined Contribution Health Benefit Plan

Just like pension benefits that can be switched from defined benefit to defined contribution type, retiree health benefits can also be switched to defined contribution. The current health benefits for retirees are largely of the defined benefit type and when state and local governments establish trust funds to prefund OPEB cost, they will also encounter risks similar to those facing defined benefit pension plans, such as investment risk and higher than expected costs in the future. One difference between these two defined benefit programs is that while government employers bear all the risks in a defined pension benefit program, they will only be partially responsible for the funding risk in a defined benefit healthcare plan, since the retirees will still have to pay a healthcare premium in most cases, whereas they do not contribute to pension plans. Despite this, changing the retirees' health benefit from defined benefit to defined contribution can still substantially reduce the future funding risk to government employers.

A health benefit plan of the defined contribution type requires the establishment of health savings accounts (HSA) for individual employees. The employer contributes a set amount or a certain percentage of an employee's payroll into the HSA, with or without employee match. When the employee retires, he can use the funds in the HSA to purchase health insurance at market rates and the government employer will no longer be responsible for his health benefits. Another funding method for defined contribution is Voluntary Employee Benefit Association (VEBA), which provides for tax-deferred contributions to a trust account. VEBAs are usually funded by "cashing out" an employee's accumulated leave upon retirement. Following are some examples of defined contribution plans established so far.

8.3.3.1 Cash Contribution

Several municipal governments have established defined contribution healthcare plans to deal with the rising healthcare costs:

- **Anaheim, California:** The City of Anaheim, California (2006) switched to a defined contribution plan to contain healthcare costs long before GASB 45. Employees hired after January 1, 1996 participate in defined contribution retiree medical plans. The city is required to make one-time contributions ranging from $3000 to $8000 per employee. The city's total contribution to the plans for the period July 1, 2005 to June 30, 2006 was $692,000. The employees can also make voluntary contributions of the value of sick or vacation hours up to certain limits into their retiree health savings accounts. These accounts are available to reimburse costs incurred after retirement for medical and other welfare benefits.
- **Oakland County, Michigan:** Even though Oakland County, Michigan (2006), has been prefunding OPEB for many years, it still found that meeting the requirement of GASB 34 will increase the cost. To contain the cost escalation in the future, the County Board of Commissioners established a new defined contribution health program, called "Retirement Health Savings Plan," in November 2005. Under this plan, employees hired after January 1, 2006 will no longer be eligible for health benefits under the old plan. Rather, the county will contribute $1300 every year (or $50 per pay period) to an employee's HSA and the employee can contribute a certain amount as well. When he retires, he can use the funds in his HSA to cover health expenses. To further limit the cost to the county government, there is also a vesting period for the benefit. The employee must work for 15 years before he is eligible for 60 percent of employer contribution and must work for 25 years before he is fully vested.
- **Orlando, Florida**: After actuarial valuation that pegged its OPEB liability at $188 million and an ARC of $22.8 million, the city of Orlando, Florida (2007) discontinued its old retiree health benefits for all employees hired after January 1, 2006 or January 1, 2007, depending on the individual bargaining group contract. Instead, the city negotiated with the International

Association of Firefighters to establish a Retirement Health Savings Program. For employees hired after July 31, 2006, the city will contribute $85 per month to the program for each employee. There is also a vesting period for city contributions. Employees will be eligible for 50 percent of the city contributions after 15 years and 100 percent after 20 years.

8.3.3.2 Converting Unused Sick Leave into Cash Contribution

Wisconsin is one of the earliest state governments to use unused sick leave to fund health benefits. The Accumulated Sick Leave Conversion Credit Program (ASLCC) was first authorized in 1972 and the Supplemental Health Insurance Conversion Credit Program (SHICC) was first authorized in 1995 (State of Wisconsin, 2007). ASLCC provides that employees retiring and beginning an immediate annuity are eligible for a termination payment of unused sick leave, deposited into an account the employee may use to purchase health insurance through the state retiree health program. The rate of conversion of unused hours to the termination account is defined in the state's various employment contracts with employees and depends on years of service, employment category, and the employees' current rate of pay. SHICC is a matching program. To be eligible for this benefit, employees must have at least 15 years of continuous state service. This program allows employees to have their ASLCC hours matched based on their years of continuous state service. These account benefits are prefunded based on an actuarially determined percentage of payroll.

In 2003, Wisconsin had an unfunded pension liability of $702 million and an unfunded sick leave retirement liability of $782 million (State of Wisconsin, 12/1/2006). In that year, the state issued a pension obligation bond of $1.35 billion at an interest rate of 5.6 percent. The proceeds were roughly split between the pension plan and the sick leave plan. The accrued health benefit liability at the end of 2005 was $1.822 billion and the program's assets were $1.805 billion (State of Wisconsin, 2006).

Utah state legislature in 2005 passed House Bill 213, Unused Sick Leave at Retirement Amendments. Employees retiring prior to January 1, 2006 were paid up to 25 percent of the value of unused accumulated sick leave and employees may receive health and life insurance coverage for up to five years or until age 65, whichever comes first (State of Utah, 2006). As of January 1, 2006, upon retirement, the employee receives 25 percent of the value of unused accumulated sick leave as a mandatory employer contribution into a 401(k) account. Health and life insurance coverage will also be phased out over the next five years. If the employee has not reached age 65 or the employee retires after the phase-out period, then that employee may exchange one day of unused accumulated sick leave earned prior to January 1, 2006, for one month of paid health and life insurance coverage. After age 65, the employee may use any remaining unused accumulated sick leave, earned prior to January 1, 2006, to exchange for spouse health insurance to age 65, or Medicare supplemental insurance for the employee or spouse. The value of any unused accumulated sick leave earned on or after January 1, 2006 will be converted to a health reimbursement account for the employee at retirement.

In 2005, the City of Arlington, Texas (2006) terminated its old retiree health benefit for employees hired after January 1, 2006 and established a defined contribution plan called the Retirement Health Savings Plan. This health benefit savings plan allows retirees to contribute and accumulate assets on a tax-free basis to pay for healthcare premiums and other health-related expenses during retirement. The plan currently allows enrolled outgoing retirees to contribute all or a portion of their accrued sick and/or vacation terminal pay into this RHS account on a pretax basis. The money can then be used by the retiree, the retiree's spouse, or other eligible dependents on a tax-free basis for qualified health expenses.

References

Aon Consulting. 2007. *State of New Jersey: Postemployment Benefits Other Than Pension Actuarial Valuation July 1, 2006.* http://www.state.nj.us/treasury/pensions/gasb-43-july2007.pdf (Accessed July 30, 2007.)

City of Anaheim, California. 2006. *Comprehensive Annual Financial Report Year Ended June 30, 2006.* http://www.anaheim.net/docs_agend/report/Anaheim_CAFR06_web.pdf (Accessed June 30, 2007.)

City of Arlington, Texas. 2006. *Workforce Services Department. Employee Benefit Guide 2007.* http://www.ci.arlington.tx.us/employment/pdf/Employee_Benefit_Guide.pdf (Accessed April 14, 2007.)

City of Gainesville, Florida. *Budget Summary FY 2006.* http://www.cityofgainesville.org/webportal/portals/omb/docs/omb_BudgetSummary.pdf (Accessed August 17, 2007.)

City of Orlando, Florida. 2007. *Comprehensive Annual Financial Report for the Fiscal Year Ended September 30, 2006.* http://cityoforlando.net/admin/accounting/PDFs/CAFR2006/CAFR.pdf (Accessed July 7, 2007.)

Edwards, Chris, and Jagadeesh Gokhale. 2006. Unfunded state and local health costs: $1.4 trillion. Cato Institute, *Tax & Budget Bulletin,* No. 40, October 2006.

Fitch Ratings. 2005. *The Not So Golden Years: Credit Implications of GASB 45.* http://www.nasra.org/resources/medical/Fitch%20OPEB%20Report.pdf (Accessed March 30, 2007.)

Gabriel, Roeder Smith, & Company. 2007. *State of California OPEB Valuation as of July 1, 2007.* http://www.sco.ca.gov/eo/pressbox/2007/05/OPEB_actuaria_report.pdf (Accessed July 30, 2007.)

McCaffrey, Shannon. 2007. Georgia Retiree Healthcare Costs Total $15 Billion. *Associated Press,* February 18, 2007.

Moody's Investors Service. 2005. *Special Comments: Other Post-employment Benefits (OPEB).* http://72.3.167.244/phpBB2/files/opeb_spl_comment_166.pdf (Accessed April 19, 2007.)

Office of the New York State Comptroller. 2007. Press release: *Testimony From First Deputy Comptroller Thomas Sanzillo Regarding GASB 45 Accounting Change.* http://www.osc.state.ny.us/press/releases/jan07/013007.htm (Accessed July 30, 2007.)

Oakland County, Michigan. 2006. *Comprehensive Annual Financial Report, for Fiscal Year ended September 30, 2005.* http://www.oakgov.com/fiscal/assets/docs/2005_cafr.pdf (Accessed March 3, 2007.)

Orange County, California. 2006. *Comprehensive Annual Financial Report For the Year Ended June 30, 2006.* http://www.ac.ocgov.com/2006CAFR/cafr06.pdf (Accessed May 3, 2007.)

Santa Clara County, California. 2006. *Comprehensive Annual Financial Statement for Fiscal Year ended June 30, 2006.* http://www.sccgov.org/SCC/docs%2FFinance%20Agency%20%28AGY%29%2Fattachments%2FCAFR%2FSCC%20FY2006%20CAFR.pdf (Accessed May 5, 2007.)

Standard & Poor's. 2004. *Reporting and Credit Implications of GASB 45 Statement on Other Postemployment Benefits.* http://www.ifpte21.org/retire/S&P-RCIofGASB45onOPEB.pdf (Accessed June 5, 2007.)

State of Connecticut, Office of the State Comptroller. 2006. Press release: *Wyman Urges Leaders To Address $21 Billion Retiree Healthcare Liability.* http://www.osc.state.ct.us/public/pressrl/2006/december29.htm (Accessed July 30, 2007.)

State of Ohio. 2006. *Comprehensive Annual Financial Report for the Fiscal Year Ended June 30, 2006.* http://obm.ohio.gov/finrep/cafr/cafr06/cafr06.pdf (Accessed April 11, 2007.)

State of Utah. 2006. *Comprehensive Annual Financial Report for the Fiscal Year Ended June 30, 2006.* http://wwwfin.state.ut.us/reports/cafr.htm (Accessed May 1, 2007.)

State of Wisconsin. 2004. *Comprehensive Annual Financial Report for Fiscal Year Ended June 30, 2004.* http://www.doa.state.wi.us/docview.asp?docid=4367&locid=3 (Accessed May 8, 2007.)

———2006. *Comprehensive Annual Financial Report For the Fiscal Year Ended June 30, 2006.* http://www.doa.state.wi.us/docview.asp?docid=5829&locid=3 (Accessed May 8, 2007.)

State of Wisconsin, Department of Administration and Department of Revenue. 12/1/2006. *State Task Force on Milwaukee County Final Report.* http://www.revenue.wi.gov/news/milwco/final.pdf (Accessed April 29, 2007.)

State of Wisconsin, Department of Employee Trust Fund. 2007. *State of Wisconsin Accumulated Sick Leave Conversion Credit Program (ASLCC) Supplemental Health Insurance Conversion Credit Program (SHICC) FACT SHEET 2007.* http://etf.wi.gov/publications/et8929.pdf (Accessed April 29, 2007.)

Wisniewski, Stan, and Lorel Wisniewski. 2004. *State Government Retiree Health Benefits: Current Status and Potential Impact of New Accounting Standards.* http://assets.aarp.org/rgcenter/health/2004_08_benefits.pdf (Accessed August 4, 2007.)

Zion, David, and Admit Varshney. 2007. *You Dropped a Bomb on Me. GASB.* Credit Suisse. http://online.wsj.com/public/resources/documents/DroppedB.pdf (Accessed September 6, 2007.)

Chapter 9

Looking Forward

State and local public pension plans have come a long way over the past century. From only limited coverage to now universal coverage among full-time government employees, state and local pension benefits are an important part of compensation for government employees and play a critical role in ensuring financial security in their retirement. The management of public pension plans has also become increasingly sophisticated over the past half century. This level of sophistication can be shown in three areas:

1. Pension funding policy has shifted from pay-as-you-go to full actuarial funding for the vast majority of state and local pension plans.
2. Pension assets are also managed more professionally through incorporation of modern investment management theories. The investment portfolio changed from an all fixed-income portfolio in the 1940s to a more diversified investment portfolio nowadays that includes equity, fixed-income, real estate, and alternative investments. While such diversification exposes state and local pension plans to more investment risk, it has brought two significant benefits to the state and local pension systems as a whole. The first is that it has led to a higher long-term investment return and a much better funding level for state and local pension systems, from about 50 percent in the 1950s to about 87 percent in 2005. The second benefit is that investment income has also significantly reduced the burden on state and local employers and employees in terms of pension contribution. In a 10-year period between 1996 and 2005, which includes the severe three-year stock market decline, total employer and employee pension contributions accounted for only one-third of the growth in pension assets, whereas investment income accounted for the remaining two-thirds.

3. Many public pension plan sponsors, because of the stock market downturn in the early 2000s, have taken steps to minimize the impact of volatility in the financial market on pension contribution and government operating budget.

With total assets of $3 trillion in 2006, state and local government pension systems are an integral part of financial management in the public sector. Overall, the majority of large state-level pension systems are well funded. After the substantial drop in overall funding ratio due to the stock market decline in the early 2000s, the average funding ratio has been on the rise again. While there are some state-level pension systems that are severely underfunded, there are also some large pension systems that are overfunded, such as the Florida Retirement System. Both the underfunded and overfunded pension systems offer important lessons for public pension management in the future.

Even though state and local pension systems have made great strides in accumulating assets to pay for future pension benefits, there are certain long-term trends that can potentially make the task of managing public pension plans more difficult in the future than in the past:

1. On the pension liability side, people are living longer. While longer life expectancy is a major achievement of modern civilization, it adds burden to the pension plans. People living longer will draw on pension benefits for a longer period of time. Each additional year of pension benefit payments adds significant cost to the pension plans.
2. On the pension asset side, the long-term investment return in the future may not be as robust as in the past. For investment return on the equity portion of the portfolio, one component of the average equity return in the past has been the increase in the overall stock market price/earning ratio, meaning investors are willing to pay a higher price for future corporate profit. As further increase in price/earning ratio will be more difficult to achieve in the future, future equity return may not be as robust as in the past. For investment return on the fixed income portion of the portfolio, a major component of the return over the past quarter century has been the decline in the overall interest rate since the early 1980s.* Given the low inflation rate and interest rate environment in the early twenty-first century, the return on fixed-income investment is also likely to be lower in the future. This expectation of lower return in the future has already led a number of pension systems to lower their assumed rate of return in recent years. A lower return will cause contributions to increase and, thus, put pressure on government plan sponsors.
3. From the broader retirement benefit perspective, while retiree health benefit is not directly related to pension benefits, it nonetheless will have an impact

* Because of the inverse relationship between yield and bond price, a decrease in interest rates led to an increase in bond price and, thus, a higher return on fixed-income investment.

on pension funding. Since the unfunded actuarial liability of Other Postemployment Benefits (OPEB) is much greater than the unfunded pension liability, the funding cost for OPEB is expected to increase substantially in the future, which will put it in direct competition with funding for pension benefit, as they all come out of the operating budget. Given the long-standing antitax attitude among the general public, raising taxes to generate additional financial resources is not a viable option for most governments, which can create pressure for public pension funding.

Given these long-term trends, an important public policy issue emerges regarding pension funding: how can we ensure financial security for public employees in their retirement without putting additional pressure on state and local government finance? In this final chapter, we will examine this policy issue from three angles. Because pension plan management involves both liability and asset management, the first two angles deal with these two areas. The third angle focuses on changes in pension benefit design that link pension liability growth to asset growth.

9.1 Managing Pension Liability Growth

On the liability side of pension management, the most important policy issue is to contain the long-term growth of pension liability. There are many approaches to containing the growth of pension liability, some more easily accomplished than others.

9.1.1 Limiting Pension Benefit Increase

As can be seen from the experience of state and local pension systems before and after the stock market downturn in the early 2000s, increasing pension benefits without identifying a permanent funding source can be as damaging to a pension system's funding status as a stock market decline. Limiting pension benefit increase is probably one of the most important factors in ensuring the long-term sustainability of public pension plans. Pension benefit increase can be treated differently depending on the funding status of a pension plan.

When a pension system is underfunded, there should be no pension benefit increase under any circumstances. Any pension benefit increase when a plan is underfunded will only increase the unfunded liability and further increase financial burden on the future generation of taxpayers. This violates the intergenerational equity principle of public finance. Such requirements can be written into a state statute or constitutional amendment. In November 1995, voters in Maine passed a constitutional amendment to such effect. The amendment requires the state to fund the unfunded actuarial liabilities of the state and teacher plans existing on June 30, 1996, over a period not to exceed 31 years beginning July 1, 1997. It also prohibits the creation of new unfunded liabilities in the state and teacher plans except those arising from experience losses, which must be funded over a period of not more than ten years (Maine State Retirement System, 2006). This constitutional amendment effectively prohibits any pension benefit increase when the pension plan is underfunded, as such increase will automatically create new unfunded liability.

While an overfunded pension plan can be permitted to grant benefit enhancements, such a benefit increase should still be dealt with cautiously. The nature of funding surplus should be analyzed. As discussed in Chapter 6, pension fund surplus is largely due to investment return greater than expected over a period of time. Because investment return over time tends to be close to the average, above average return is most likely to be followed by below average return. Therefore, temporary pension surplus should not be used for permanent pension benefit increase, unless there is reason to believe that investment return is going to be permanently higher in the future. A more prudent usage of temporary funding surplus is to require at least part of the surplus be kept in a pension fund reserve, as is done in Florida.

If pension benefit enhancement is allowed after a certain portion of the surplus is preserved, then the plan sponsor should be required by law to fully disclose the information on the cost and the funding source associated with such enhancement before it can be enacted:

1. For the cost, the plan sponsor should disclose the total cost or the net increase in present value of future pension liabilities due to the enhancement. Ideally, the total cost should also be broken into two parts, the part attributable to retroactively granting an attributable increase to services already provided and the part to services yet to be provided.
2. Then the plan sponsor should be required to disclose how this total net cost would be funded. There are only three sources to cover such a cost: the temporary pension surplus, and employer and employee pension contribution increases. If the pension surplus (minus the preserve) is large enough to cover the entire enhancement cost, there is no need to increase the employer and employee pension contribution rates. If the total cost is greater than the surplus, then the pension contribution rate has to go up. In this case, the plan sponsor needs to disclose when and by how much the pension contribution rate will have to go up and how the rate increase will be distributed between the employer and employee, and the actual dollar amount associated with the rate increase over the next five to ten years. To be more thorough, it should also be accompanied by a projected operating budget growth over the next five years to show whether the operating budget can afford such a rate increase in the future. A stricter funding requirement is that at least the enhancement cost due to past services should be fully funded at the time when the benefit enhancement legislation is passed.

Another way to put strict limitation on pension benefit increase is to take the decision out of the elected officials' hands and put it squarely in the hands of voters/taxpayers. This way, it will force elected officials to explain very clearly to voters how much the enhancement will cost and how it will be funded, and whether such enhancement is justifiable. Then it will be up to the voters to decide whether they want to pay for such an enhancement. Requirement of voter approval does not automatically preclude a benefit increase. The city of San Francisco has a century-old provision in its city charter that requires voters to approve any pension benefit increase to city employees. In the ten years prior to 2006, residents there approved at least four modest benefit increases (Halper, 2006). Partly because of this strict limitation, the city and county of San Francisco Employees' Retirement System (2006) had assets valued at $12.7 billion and a funding ratio of 108 percent at the end of fiscal year 2006, one of the highest in the country.

In the aftermath of the San Diego city pension crisis, largely as a result of several pension benefit enhancements, voters in San Diego approved an amendment to the city charter in 2006: in a fifteen-year period beginning in January 2007, any pension benefit increase to city employees has to be approved by the voters.

Whatever limitation on pension benefit increase is set in place, the main purpose is to make sure that the funding of pension benefits is transparent and affordable, and they are paid for when they are earned as opposed to transferring the bulk of the cost to the future generation.

9.1.2 Increasing Retirement Age

Another approach to containing long-term growth in pension liability is to increase the normal retirement age for regular employees in the public sector. The life expectancy continues to increase gradually for both men and women in the United States, both at birth and at the age of 65. Table 9.1 shows the historical trend of years a man or woman is expected to live at the age of 65 (Board of Trustees, 2007).

A man collects pension benefits for 3.4 more years in 2006 than in 1970, and by 2035, he is expected to collect pension benefits for almost another two years. The trend is similar for a woman, although the increase in life expectancy is not as large as that for a man. If the normal retirement age is not increased, it will lead to a higher pension liability. Such an increase in pension liability can be funded in two ways. If this higher cost is funded when the employee is still working, assuming the actuarial assumption is changed to reflect the longer life expectancy, it will lead to higher employer and/or employee pension contributions. If the higher cost due to longer expectancy is funded after the employee is retired, then it has to be funded

Table 9.1 Years of Life Remaining at Age 65

Year	*1970*	*1980*	*1990*	*2000*	*2006*	*2035*
Male	13.1	14	15.1	15.9	16.5	18.3
Female	17.1	18.4	19.1	19	19.1	20.5

Source: The 2007 Annual Report of the Board of Trustees of the Federal OASDI.

entirely by higher employer pension contribution and/or higher pension contribution for future employees.

One way to reduce the cost related to longer life expectancy is to increase the normal retirement age. An additional year of employment will bring in additional pension contribution from the employee, an additional year for pension asset to earn income, and one less year for pension benefit withdrawal. These three aspects serve to reduce the cost to the government pension plan sponsor. Increased life expectancy is the reason why the age at which unreduced Social Security benefits can be collected has been increased from 65 gradually to 67 for people born after 1960.

By comparing the Wisconsin survey findings of normal retirement age requirement for the 83 large state-level pension systems in 1982 and 2004, there has been no increase in normal retirement age. In many cases, the normal retirement age has actually been decreased rather than increased over this period. As one example of this reduction in normal retirement age, there were 12 plans that used age 65 as the exclusive requirement for normal retirement benefits. By 2004, only two (both in Washington State) of the original 12 plans still used such a requirement. Other plans adopted different criteria, usually a combination of age and years of service that effectively reduce the normal retirement age. Besides, many plans allow for normal retirement after 30 years of service without regard to the age. This is in sharp contrast to the normal retirement age requirement in the private sector. Table 9.2 shows the percentage of employees participating in medium and large private pension plans with various age requirements for normal pension benefits in 1995 and 2005, and those percentages for state and local pension plans in 1998 (U.S. Department of Labor, 1998; 2000; 2007).

Four observations can be made of this table. First, age 65 is the normal retirement age for most employees in the private sector. Second, close to 80 percent of employees in the private sector can retire at age 62 or older with unreduced pension benefits. As discussed in Chapter 2, 62 is the earliest age a person can collect Social Security benefits, although at a reduced rate. Third, over the ten-year period between 1995 and 2005, there was a noticeable increase in private plans that set the

Table 9.2 Age Requirement for Normal Pension Benefits (%)

Year	*Pension Plan*	*Age 65*	*Age 62*	*Age 60*	*Age 55*
1995	Median and large private pension plans	48	26	8	4
2005	Median and large private pension plans	62	15	10	1
1998	State and local pension plans	9	4	10	

Source: Department of Labor, Bureau of Labor Statistics, 2007, 2000, 1995, Washington, D.C.

normal retirement age at 65. Fourth, the normal retirement age overall was much lower in the public sector in 1998.

In line with the increased life expectancy and the pension cost associated with it, normal retirement age for new regular public sector employees should be increased to 65, so as to tie public pension benefits with Social Security benefits. Once the normal retirement age is increased, the early retirement age should be increased accordingly. So far, at least two major state-level pension systems, both with a funding ratio substantially below national average, increased the normal retirement age in recent years to reduce future pension liability growth:

1. In 2007, Kansas passed Senate Bill 362 to reform the Kansas Public Employees Retirement System (KPERS), whose funding ratio was 75 percent. Among the many provisions of the bill, one provision increased the retirement age. Under the old rule, KPERS members could retire under Rule-85, or at age 65 with no years of service requirement. Under the new rule, for employees hired after July 1, 2009, the normal retirement age will be 65 with five years of service, or age 60 with thirty years of service (State of Kansas).
2. In Rhode Island, the state legislature in 2005 passed a pension reform package to reduce future pension cost. The Employees Retirement System of Rhode Island was only 60 percent funded as of June 30, 2004 (Employees Retirement System of Rhode Island, 2005). For state employees and teachers with less than ten years of service as of July 1, 2005 and all new employees hired after July 1, 2005, the pension reform increases the minimum retirement age eligibility from age 60 with ten years of service (or any age after twenty-eight years of service) to age 65 with ten years of service, or age 59 with twenty-nine years of service (State of Rhode Island).

9.1.3 Pension Benefit Level

The increase in normal retirement age to 65 should also be combined with a cap on pension benefit level that is fair to public employees. Being fair to public employees means that the benefit level should ensure that the quality of life be maintained after retirement.

As discussed in Chapter 2, about 85 percent of preretirement income is needed to maintain the quality of life. By increasing normal retirement age to 65, then pension benefits can be more easily integrated with Social Security benefits to ensure the quality of life. Since Social Security benefits will replace about 30 percent of income at age 65 for an average public sector retiree, as discussed in Chapter 2, then limiting pension benefit to somewhere between 60 and 75 percent of preretirement income can ensure the quality of life. Based on the calculation in Chapter 2, the replacement ratio from public pension benefit is found to be 55 percent. However, due to the final salary averaging and assuming a 5 percent salary growth, the real replacement ratio should be around 60 rather than 55 percent when using final average salary in the benefit formula. Thus, at 60 percent, the combined pension and Social Security benefits add up to about 85 percent of actual preretirement income.

While pension benefit equal to 60 percent of preretirement income is the minimum needed to maintain living standards in retirement, the maximum pension benefit should not exceed 75 percent of preretirement income. At 75 percent (equivalent to 70 percent replacement ratio of preretirement income), the combined benefits will be equal or very close to 100 percent of preretirement income for the vast majority of public employees. As an example, Florida state law requires that combined pension and Social Security benefits cannot exceed 100 percent of preretirement income. Even though Florida does not set a limit on pension benefits, this law sets an implicit upper limit on the pension benefit level. If pension benefit exceeds 75 percent of final average salary at the age of 65, then it is very likely that combined pension and Social Security benefits will exceed 100 percent of preretirement income, calling into question the purpose of granting such a pension benefit in the public sector. If the preretirement income is too low and needs to be compensated by higher pension benefit, then the less transparent nature of pension cost has the potential to mask the true cost of public service provision and lead to pension underfunding and intergenerational inequity, the theme of financial illusion theory discussed in Chapter 6.

Limiting pension benefits to 75 percent or less of final salary is not unprecedented in the public sector. As discussed in Chapter 2, quite a few pension systems limit the pension benefit to 75 percent or less of final average salary. A cap on pension benefit level will be most effective when there is a corresponding change in the benefit multiplier. The benefit multiplier should be set at a level that the cap on pension benefits can be reached only at the retirement age of 65 after a full career in the public sector. A full career at age 65 should be somewhere between thirty-five and forty years, meaning the pension benefit formula should be set in such a way that the 60 percent or 75 percent cap on pension benefits can only be reached after thirty-five to forty years of service. Any employee who wants to retire before the age of 65, therefore, needs to receive reduced pension benefits, as in the case of Social Security benefits.

As a more recent example of reducing the maximum pension benefit to control costs, in the pension reform package passed in Rhode Island in 2005, another important provision is to decrease maximum benefits from 80 percent of the highest three years' salary at thirty-five years of service to 75 percent at thirty-eight years of service. In conjunction with the lower cap, the benefit multiplier has also been reduced. Prior to the reform, it was 1.7 percent for the first ten years, 1.9 percent for the second ten years, 3.0 percent for years twenty-one through thirty-four, and 2.0 percent for years over that. After the reform, the multiplier is 1.6 percent for the first ten years, 1.8 percent for the second ten years, 2.0 percent for years twenty-one through twenty-five, 2.25 percent for years twenty-six through thirty, 2.5 percent for years thirty-one through thirty-seven, and 2.25 percent for year thirty-eight.

In addition to controlling pension liability growth, a cap on pension benefit level has several other advantages. First, it presents an image to the public of the fairness of public pension plans. Second, the effect of any future pension benefit increase will most likely be more muted as there is an upper limit on how high the pension benefit can go. Third, it can also limit the negative effect on the pension system of any potential pension benefit abuse, more commonly known as "pension spiking."

9.1.4 Containing the Negative Effect of Spiking

Pension spiking means a much larger increase than normal in income in the final years to obtain a much larger pension benefit for life. Such a larger salary increase can be due to overtime, bonus, sick leave, and vacation time buyback, or larger than normal salary increases in the final years for the same job. From time to time, there are media reports about some employees collecting pensions greater than their pre-retirement income. An investigation in 1998 by the State of New Jersey Commission of Investigation unearthed many abusive practices among local governments that substantially boosted the final pension benefit (State of New Jersey, 1998). As pension liability is based on projected final salary, a much higher actual final salary than projected will substantially increase the pension liability. While pension spiking is by no means prevalent in the public sector, it can still cause shortage in funding. More importantly, it can damage the public's confidence in the integrity of public pension benefit program.

While an overall cap on pension benefit level, as suggested above, can limit the negative effect of pension spiking, another effective way to deal with spiking is to limit the increase in total salary increases in the final years. The limit can be either indexed to a benchmark or set to a constant rate. As an example of indexing, in its report on pension benefits review, the New Jersey Benefits Review Task Force recommended limiting the salary growth in the final five years to the average salary growth of the entire plan (State of New Jersey, 2005). As an example of a constant rate, in the Kansas Senate Bill 362 that was passed in 2007 to reduce pension costs, it also contains another provision that sets a cap of 7.5 percent on pay increases in the five highest years for determining benefits.

9.1.5 Early Retirement Incentive Program

Early retirement incentive programs should be avoided in the public sector because its chief objective of cost savings is difficult to achieve in the public sector. While the cost, in terms of the increased pension liability, can be calculated with relative accuracy, the savings of such a program is hard to project. As discussed in Chapter 2, most studies have found the savings is always much less than projected and also much less than the increase in pension costs in the long run. Because such programs only result in short-term budget savings in personnel costs at the expense of much greater increases in the long-term pension liability, it should be avoided.

9.1.6 *Pension Obligation Bond*

If a pension plan has an unfunded liability, issuing a pension obligation bond can still be a viable option to reduce the long-term amortization cost, as long as it is used prudently. If the pension obligation bond (POB) can be issued at an interest rate that is below the pension plan's assumed rate of return, such prudence can be achieved in several ways:

1. A law can require the government not to increase pension benefits when the bond is still outstanding, as tried by the West Virginia legislature in its failed attempt to issue a POB in 2005.
2. Government has to meet its actuarially required contribution every year when the bond is outstanding.
3. The proceeds of the POB can only be used to pay down unfunded liability and not for offsetting future pension contribution.
4. The debt service payment on the POB should not be backloaded, meaning the debt service payment schedule should not be structured in such a way that the bulk of the principal payment is pushed into the distant future. The debt service payment can either be level dollar or level percentage of payroll, which is similar to the methods used to determine the amortization cost on unfunded liability with a level percentage method. This ensures that the savings from the POB are spread out more evenly in the future.

9.2 Managing Pension Asset Growth

Containing pension liability growth is only part of the answer for ensuring long-term sustainability of public pension benefit programs. The other part is to ensure adequate pension asset growth in the future, entailing both adequate pension contribution and investment return.

9.2.1 *Pension Funding Policy*

The establishment of a strict pension funding policy is the key to ensuring adequate pension contribution. Even without any federal regulation, the majority of state-level pension systems have made paying the actuarially required contribution their funding policy, which is also the funding policy required under ERISA (Employee Retirement Income Security Act). Only a handful of large state-level pension systems have seen their actuarially required contributions chronically underfunded by the government plan sponsors in the past. Such chronic underfunding led four state-level pension systems to have a funding ratio below 60 percent in 2004: Connecticut State Employees Retirement System at 57 percent, Illinois State Employees Retirement System at 54 percent, Oklahoma Teachers Retirement System at 47 percent, and West Virginia Teachers Retirement System at 22 percent.*

Encouragingly, all four states have passed laws to either gradually increase pension contributions or meet the actuarially required contribution, so as to increase the pension-funding ratio in the future. As pointed out in Chapter 6, Illinois and West Virginia passed laws in the 1990s to gradually increase employer pension contribution and the funding ratio to 90 percent over a 40-year period. West Virginia has even gone beyond what is required by law in recent years by depositing a large portion of budget surplus into the retirement system.

In Oklahoma, the state government passed SB 357 in 2007 to increase employer contributions to the Teachers' Retirement System (TRS). Under the measure, state contributions to the system would increase from 7.85 percent in 2007 to 9.5 percent in 2010.† The law also contains a provision to ensure state contributions to TRS will be held harmless in periods of economic downturn. If the funding is put in each year, the bill would put the state on track to fund the TRS at 80 percent by 2026 (Oklahoma Teachers Retirement System).

In Connecticut, Governor Jodi Rell signed into law Public Act 07-186, *An Act Concerning Adequate Funding of the Teachers' Retirement System*, authorizing the issuance of a pension obligation bond (POB) for up to $2 billion with a maximum term of thirty years to fund the unfunded liability of TRS. For each year the bond is outstanding, the law establishes an automatic General Fund appropriation for the state's actuarially required amount, which is also certified by the TRS and state comptroller. The law also promises that the general assembly will not pass any law to diminish the state's required contribution until the bonds are paid off unless the governor declares an emergency, three-fifths of the legislators approve the reduction, and the TRS funding ratio cannot be reduced by more than 5 percent due to the reduction or below the funding ratio immediately after the POB sale, whichever is greater.‡ After two decades of underfunding pension contribution, the state of Connecticut fully funded its pension obligation from fiscal years 2006 through 2009 (State of Connecticut, 1/15/2008). In January 2008, Governor Rell announced the

---

* Another large state-level pension system that had a funding ratio below 60 percent in 2004 is Indiana Teachers Retirement Fund, as discussed in Chapter 4. The low funding ratio is due to a pay-as-you-go funding method for plan members hired before 1995.

† The full text of the bill is available at Oklahoma Secretary of State Web site at http://www.sos.state.ok.us/documents/Legislation/51st/2007/1R/SB/357.pdf (Accessed July 27, 2007.)

‡ The full text of the law can be found at Connecticut state legislature's Web site at http://www.cga.ct.gov/2007/ACT/PA/2007PA-00186-R00HB-06141-PA.htm (Accessed September 8, 2007.)

sale of $2 billion in POB to help close a $6.9 billion unfunded liability.* The sale was approved by the State Bond Commission in the same month.

These major changes in funding policy made by states with some of the worst funded pension systems indicate that state governments that have not fully funded the actuarially required amount have realized that pension obligation can no longer be neglected and urgent action is needed to bring the pension systems to healthy levels.

9.2.2 Smooth Pension Contribution Rate

For those plan sponsors who have paid the actuarially required contribution consistently, a more important funding policy issue is how to avoid the ups and downs in the contribution rate, especially during times when the pension fund is temporarily overfunded. Such volatility is damaging to both the pension plan and the government's operating budget. Even though it will be many years before many pension systems will become overfunded again, it is still advisable for plan sponsors to design funding policy in the context of overfunding to avoid volatility. Again, it is also encouraging to see that some state governments have learned the lesson after the 2000 stock decline by designing policies to smooth out pension contribution, such as the minimum pension contribution law in New York State. A more detailed discussion on this subject can be found in section 6.4 of Chapter 6.

9.2.3 Investment Return

Compared to pension contribution, public pension systems have less control over what investment return they will get and the risk they will face. The shift from fixed-income to equity since the 1950s has benefited public pension systems, but also has brought some volatility in asset valuation. To reduce the volatility from the equity market and to boost future investment return given the expectation of potential lower return in the equity and fixed-income markets in the future, there has been another gradual shift in asset allocation among many public pension systems, by reducing the allocation to traditional investment and increasing the allocation to alternative investments, such as private equity, hedge funds, and real estate. Alternative investments have lower correlation with traditional investment and potentially higher return. Following are asset allocation policy targets for real estate and alternative investments for some large state public pension systems in 2006:†

* Ibid

† Unless otherwise indicated, these policy targets are from the respective retirement systems' annual financial report for fiscal year 2006.

- California Public Employees Retirement System: 14 percent
- California State Teachers Retirement System: 12 percent
- Oregon Public Employees Retirement System: 20 percent
- Pennsylvania Public Employees Retirement System: 18 percent
- Pennsylvania State Employee Retirement System: 20 percent
- Washington State Retirement Comingled Trust Fund: 28 percent (Washington State Investment Board, 2006).

Those pension systems that have devoted a significant portion of their portfolio to alternative investments have also benefited since 2000 in terms of higher return and reduced volatility. However, given the relative short history of public pension systems investing a significant portion of their portfolios in the alternative asset class, it still remains to be seen whether such a strategic shift in asset allocation can help them meet or beat the assumed rate of return with reduced volatility in the long run.

9.3 Changing Pension Benefit Design

While the containing of pension liability growth and the achieving of adequate pension asset growth have been discussed separately so far, pension liability growth can also be linked to pension asset growth to ensure long-term sustainability of public pension benefit program. As discussed in Chapter 7, the plan sponsor bears all the funding risk in a defined benefit pension plan, whereas plan members shoulder all the risk in a defined contribution pension plan. However, there are pension benefit designs that allow plan sponsors and members to share the funding risk.

9.3.1 *Hybrid Plan*

One way to share funding risk is to establish a hybrid plan. There are two basic types of hybrid plan: cash balance plan or a pension plan with both defined benefit and defined contribution components, as discussed in Chapter 7. In a cash balance plan, the employer guarantees a lower rate of return and part of excess return over that is also passed onto employees. In a plan with both defined benefit and defined contribution components, employer contribution typically goes into a defined benefit plan and employee contributions go into a defined contribution plan. In this case, the employer also guarantees a lower level of pension benefit and the employee is responsible for part of his retirement benefit, which is dependent on the investment performance. To further limit the risk to employees in the defined contribution component, employers can put employees in a default well-diversified investment fund, such as a life cycle fund, unless they opt out.

While the risk is shared somewhat differently within these two plans, both of them essentially guarantee a somewhat lower level of benefit compared to a traditional defined benefit plan, and then let the financial market determine how much higher the overall pension benefit level can go. This will limit the employer's exposure to funding risk without overly compromising the employee's financial security in retirement.

9.3.2 Cost of Living Adjustment (COLA)

A less fundamental way of changing the pension benefit design that allows the sharing of risk between employer and employees has to do with cost of living adjustment (COLA). COLA is an important benefit that protects retirees against the erosion of purchasing power due to inflation. It also adds a significant cost to the employer as well because such adjustment is permanent. As discussed in Chapter 2, there are two basic types of COLA adjustment: a fixed rate adjustment and an ad hoc adjustment. For the fixed rate adjustment, the rate is usually set at 3 percent. For an ad hoc adjustment, it is either a true ad hoc with no conditions attached or based on the investment return, meaning the adjustment is provided only when the investment return is above a certain level for a period of time. The sharing of funding risk can be done through the combination of the fixed rate and the ad hoc methods based on investment return, very similar to the sharing of risk methods discussed above. This sharing entails setting the fixed rate to a lower percentage, say, half of the current fixed rate for most pension systems, and then letting the financial market determine the additional adjustment that can be provided. This way employees will be guaranteed some level of protection against inflation, and the impact on the pension asset value can also be somewhat blunted during a downturn in the financial market. In fiscal year 2007, the Montana state legislature passed a law to change the guaranteed increase. For employees hired before July 1, 2007, the increase is 3 percent; for employees hired after that, the increase was reduced by half to 1.5 percent (Montana Public Employees Retirement System, 2007).

9.4 Conclusion

The stock market decline of 2000 to 2002 was a sober wake-up call to many public retirement systems, even for those who have consistently fully funded the actuarially required amount. As shown in this book, many governments have passed laws since then to either make future pension contribution more consistent or increase future pension contribution in order to make up for past contribution deficiency and bring up the funding ratio. While state governments have made great strides on the pension asset side, still more can be done on the pension liability side. With some of the suggestions in this chapter on containing the growth of pension liability in the future and with the stepped-up efforts by many state governments to contribute, the public sector pension systems will be able to provide fair pension benefits to employees for decades to come at a reasonable cost to the taxpayers.

References

Board of Trustees, Federal Old-Age and Survivors Insurance and Federal Disability Insurance Trust Funds. 2007. The 2007 Annual Report of the Board of Trustees of the Federal Old-Age and Survivors Insurance and Federal Disability Insurance Trust Funds (Washington, DC: U.S. Government Printing Office).

City and County of San Francisco Employees' Retirement System. 2006. City and County of San Francisco Employees' Retirement System Annual Report, for Fiscal Year ended June 30, 2006. http://www.sfgov.org/site/uploadedfiles/sfers/2006_Annual_Report_Draft.pdf (Accessed April 9, 2007.)

Employees Retirement System of Rhode Island. 2005. Annual Financial Report for the Fiscal Year Ended June 30, 2005. http://www.ersri.org/public/documentation/AR-05.pdf (Accessed May 2, 2007.)

Halper, Evan. 2006. S.F. is Seen as a Model of – Restraint? Los Angeles Times, December 30, 2006, B1.

Maine State Retirement System. 2006. Comprehensive Annual Financial Report, for the Fiscal Year ended June 30, 2006. http://www.msrs.org/pdfs/publications/06ar.pdf (Accessed May 9, 2007.)

Montana Public Employees' Retirement System. 2007. Defined Benefit Retirement Plan Basics. http://mpera.mt.gov/docs/DBBasics.pdf (Accessed 11/7/2007.)

Oklahoma Teachers Retirement System, *Legislative Alert*. http://www.ok.gov/TRS/documents/LEGISLATIVE%20ALERT%20SB357.pdf (Accessed June 19, 2007.)

State of Connecticut, Office of the Governor. 1/15/2008. Press release*: Governor Rell Authorizes $2 Billion in Bonds to Reduce Unfunded Liability in Teachers' Retirement System*. http://www.ct.gov/governorrell/cwp/view.asp?Q=404126&A=3293 (Accessed January 17, 2008.)

State of Kansas, Office of the Governor. 4/20/2007. News release. *Providing benefits to KPERS retirees and making other changes to the retirement system*. http://www.governor.ks.gov/news/NewsRelease/2007/nr-07-0420a.htm (Accessed June 18, 2007.)

State of New Jersey, Benefits Review Task Force. 2005. The Report of the Benefits Review Task Force to Acting Governor Richard J. Codey. http://www.njspba.com/Documents/final_report.pdf (Accessed March 13, 2007.)

State of New Jersey, Commission of Investigation. 1998. Pension and Benefit Abuses. http://www.state.nj.us/sci/pdf/pensions1.pdf Accessed 6/7/2007.)

State of Rhode Island, Department of Administration. *Fiscal Fitness Project Information*. http://www.fiscalfitness.ri.gov/documents/project_portfolio.pdf (Accessed June 16, 2007.)

U.S. Department of Labor, Bureau of Labor Statistics. 2007. National Compensation Survey: Employee Benefits in Private Industry in the United States, 2005, Washington, DC: Government Printing Office.

———2000. Employee Benefits in State and Local Governments, 1998, Washington, DC: Government Printing Office.

———1998. National Compensation Survey: Employee Benefits in Medium and Large Private Establishments, 1995, Washington, DC: Government Printing Office.

Washington State Investment Board. 2006. Twenty-Fifth Annual Report. http://www.sib.wa.gov/financial/pdfs/annual/ar06.pdf (Accessed August 6, 2007.)

Index

A

B

D

H

I

M

N

O

P

R

S

T

U

V

W

Y

Z